Gardener's Guide to
TROPICAL PLANTS

First published in 2012 by Cool Springs Press, an imprint of the Quayside Publishing Group, 400 North First Avenue, Suite 300, Minneapolis, MN 55401 USA.

The information in this book is true and complete to the best of our knowledge. All recommendations are made without any guarantee on the part of the author or Publisher, who also disclaims any liability incurred in connection with the use of this data or specific details.

Cool Springs Press titles are also available at discounts in bulk quantity for industrial or sales-promotional use. For details write to Special Sales Manager at Cool Springs Press, 400 North First Avenue, Suite 300, Minneapolis, MN 55401 USA. To find out more about our books, visit us online at www.coolspringspress.com.

ISBN-13: 978–1–59186–532–2

Library of Congress Cataloging-in-Publication Data

Neal, Nellie.
    Gardener's guide to tropical plants : cool ways to add hot colors, bold foliage, and striking textures / Nellie Neal.
       p. cm.
  Includes bibliographical references and index.
  ISBN 978-1-59186-533-9 (pbk.)
  1. Tropical plants. 2. Plants, Ornamental. I. Title.

  SB405.N37 2012
  635.9'523--dc23

                         2011035909

President/CEO: Ken Fund
Group Publisher: Bryan Trandem
Associate Publisher: Mark Johanson
Acquisitions Editor: Billie Brownell
Design Manager: Brad Springer
Production Manager: Hollie Kilroy
Production: S. E. Anderson
Horticultural Editor: Steve Asbell

Printed in China

10 9 8 7 6 5 3 4 2 1

Gardener's Guide to
TROPICAL PLANTS

Cool Ways to Add Hot Colors, Bold Foliage, and Striking Textures

Cool Springs Press

Growing Successful Gardeners™

MINNEAPOLIS, MINNESOTA

Dedication

To my children, who grew up to be people I would like even if we weren't related. Thank you for loving words. As always, thank you, Dave.

Acknowledgments

My passion for tropical plants comes from deep in my soul, but most of my knowledge of how to grow them comes from time spent in the greenhouses at LSU with Dr. Ed O'Rourke and Dr. Carlos Smith. No one ever had better mentors and friends. My curiosity and the beauty of these plants led me to study more tropicals, and hundreds of questions from gardeners steer me to the rest, day by day. There is a horde of people who helped with this book by enthusiastically offering ideas and feedback. You know I owe you, and I know you will collect.

My heartfelt thanks to the unsinkable Billie Brownell, Editor Extraordinaire, for her calm hand on the tiller. She and her team at Cool Springs Press turn words into books and that is the greatest gift a writer can get. Thanks to Horticulture Editor Steve Asbell for wise counsel and taxonomic wizardry, and to Glenn Stokes, who has wooed many to grow tropical plants with his catalogs, website, and remarkable photography. This book was originally the brainchild of Cindy Games and the late Roger Waynick, to whom I owe a debt of thanks for its genesis.

Contents

Welcome to Growing Tropical Plants

When travelers dream of the tropics, there are palm trees, sandy beaches, and rum drinks for everyone. Gardeners, though, see much more than that. We see a technicolor display of colorful flowers, crazily painted leaves, and huge shapes that can dominate any landscape. We see the complexity of the blooms, long tubes with big disks, and bracts that have bodacious ways to attract the proper pollinators. Their colors span the spectrum and share a depth of hue—even in light shades—that is rare outside frost-free zones.

The lines they create are a wild ride for the eyes. They soar, curve, twist, and excite passions with scents that make us look their way. The patterned leaves on tropical plants can be neatly striped like a barber pole or seemingly random patches of color thrown together with no apparent reason. Tropical plants may be tall or short, but they are never wimpy; they present strong outlines and textures that cannot be ignored.

Universal Pleasures

When gardeners see the variety, diversity, and drama offered by tropical plants, it is natural to want to grow them. If we live in tropical zones, each discovery is like a new favorite ice-cream flavor. We collect both those that thrive where we live and those from other tropical microclimates that require different conditions. Gardeners in the subtropic zones have great latitude in the tropical plants they can grow. Some of these plants are semihardy and may become perennial plants where frosts are few and far between; others can be grown as summer annuals and container specimens alone or in mixed pots. The subtropics offer numerous ways to keep tender plants alive

7

A tropical landscape is a technicolor display of colorful flowers, painted leaves, and bold shapes.

indoors and over the winter. For too long, temperate-zone gardeners could find only a limited number of tropical plants at their favorite garden centers, but that has changed drastically. The overall availability of exotic plants has expanded by leaps and bounds in recent years as both local and mail order sources have taken advantage of the growing market. Whether grown as summer bedding plants or for long life in a container, tropicals can easily find a place in temperate zone gardens.

This book is designed to enable gardeners everywhere to better enjoy tropical plants. It will immerse you in their seductive charms and reveal their origins. These plants carry a mysterious aura with them wherever they go, but growing them will be quite straightforward with the help of this book. If you like tropical plants but are not sure why, their appeal is explained in terms of emotional allure and landscape impact, since both contribute to the totality of tropical beauty. You will learn about the features of particular plants so you can make smart choices, and you'll find out how to care for these plants during each season. The discussion includes tropical plant growing both inside and outdoors, with simple propagation suggestions and details of likely challenges, including common pests. The tropical plants profiled here represent more than 130 plants that can be grown in and out of frost-free zones in garden beds and containers. Taken as a group, they represent a gardener's musical riffs on the theme of tropical tone. The individual profiles offer suggestions about where and how to grow that plant, which ones to plant in combination to create companionable vignettes, and notes other related plants that would be worth a look. Elsewhere, the book puts design tools at your fingertips to help you incorporate tropicals into your garden or start an oasis from scratch. Plant lists for various conditions and qualities are provided at the end of the book, along with a section of resources you will find valuable in building your tropical collection.

Family Ties

The tropics is a place, but it is also a state of mind. It is an obi, a wide sash around the waist of this big blue marble we call Earth. The belt spreads to 23 degrees 27 minutes north and south of the equator, a wide swath as diverse as the scores of cultures that arose there. Perhaps more than any other area of the world, the food, music, and plants of the tropics are embraced everywhere, even by people who have not yet visited the area. We grow tropical plants for the way they make us feel as surely as we feel the rhythm of *son Cubano* music in our hips without ever having visited Havana. Our instinctive reactions begin with feelings of warmth—literally, as the sun beats down and every cuisine includes peppers and spice. But our responses are also emotional and heartfelt, sometimes unexpectedly so. The tropics draw us to bask, to slow our pace and radiate good will as we ponder waves and undulating vines dripping with blooms. Those who live in the frost-free zones inhabit a different world, one where the sun beams directly down on our lives as if in affirmation of our good fortune. Everywhere else, the sun's rays come in at an angle, with less intensity year-round. Here is the crucible of weather, where the trade winds crash together. The trade winds blow from the southeast to the northwest below the equator, and from the northeast to southwest above it. Over the

equator—only in the tropics—the collision of these winds cook up weather patterns that travel the world. These forces are emblematic of the magnetism and power of the tropical belt, as are the tropical rainforests, Nature's most efficient air-cleaning machines.

Put it all together and you get a blazing, bright, humid, languid, truly exotic place where perfumed fragrances waft day and night. The plants are flamboyant and gregarious, secretive and exotic by turns, each evolving in response to these climatic limitations with wildly different growth habits. Those plants that share certain traits are said to be related and are grouped into families to reflect those similarities. These taxonomic designations can be useful because knowing a plant's family ties helps to organize them in your

Tropical plants are often extravagantly exotic, both in colors and forms.

head. For example, most Euphorbiaceae family members ooze a latex sap when cut. Many of us have seen this sap when a branch breaks off a Christmas poinsettia. Put those two facts together and look it up—poinsettia is a member of the Euphorbiaceae family as well. This insight leads to others, as when you are instructed to let succulent and oozy stems rest—heal over or form callus—before rooting them. Some of the plants that need this treatment must be on the Euphorbiaceae family tree, too, and poinsettia cuttings must need to heal over. They are, and they do. This is an example of how knowing the taxonomy of your plants can help you to care for them properly.

The plants included in this book hail from many plant families, those taxonomic groups determined by shared traits. More than a dozen popular families are well represented here, each united by very specific botanic structural analysis but with other commonalities handy to know when growing them. There are members of the Acanthaceae family, marked by thinner leaves than most tropical plants. These leaves are strikingly beautiful up close, whether green or multicolored. Among this family are firecracker flower, nerve plant, and sanchezia, a diverse group with shared roots. Upright elephant ear, dumb cane, and caladium are examples of the Araceae family members whose flowers reveal their kinship even though all are grown primarily for their outrageous leaves. Their flowers are rubbery spikes with a fleshy bract cupped like Dracula's cape around a long nose of tiny flowers. There are classic tropical palms in this book that belong to the Arecaceae family. Asteraceae, also called Compositae, is the daisy family, known for its flowers made of disk and ray florets. You can see them in gerber daisy, Mexican flame vine, and other plants included here. Members of the

A lush tropical garden invites guests to linger and enjoy its ambience.

A garden in Costa Rica

Begoniaceae, Bromeliaceae, Orchidaceae, and Cactaceae families made it into the collection and are familiar to most gardeners, but the Euphorbiaceae family plants are a more diverse group than most imagine. Yes, crown of thorns is a member, but so are copperleaf and chenille plant. It is interesting to know that Fabaceae is the bean family, seen here in plants that do, in fact, produce beans as their seedpods. Members of the Lamiaceae, or mint, family run the gamut of colors and flower forms, but all have square stems, perhaps best seen in lion's ear and coleus. Last, but not least, Solanaceae is well represented in this collection, but none resemble a tomato or eggplant unless you examine their flower parts very closely. Here you will encounter the romance of night-blooming jasmine and the oddity of a plant called yesterday, today, and tomorrow.

The plant families dance around the world, one member here and another there, linked by scientific notation of their distinctions but unknown to one another in nature. For example, the Rubiaceae family includes woody plants such as Malaysia's yellow gardenia and flame of the woods from Sri Lanka. Likewise, orchids are native to tropical climates in both Asia and the Americas. Our joy as gardeners is to bring them together in the endless combinations, like the steps in a salsa variation or the storytelling arms of hula dancers.

The Big Three

The vast swath of Earth known as the tropics comprises most of the world's biomes, from sandy beaches to snowcapped peaks. Plants in the tropical category for this book primarily derive from rainforests, savannas, and deserts—what I'll call the Big Three habitats of the tropics. Elevation, rainfall, and soil conditions vary widely among these habitats, and plants have evolved to suit each distinctly.

The rainforests average more than 100 inches of rain annually where heat and humidity build on summer afternoons to produce cumulus clouds and showers. Winter is wet, too, in this region that spans the equator from 10 degrees S to 25 degrees N. The rainforests' average temperature of 80 degrees Fahrenheit matches the average humidity of 80 percent. The rainforests circle the globe: the Amazon and Congo basins and the East Indies from Sumatra to New Guinea. When you grow plants native to these regions, you create a damp, deep green world of shadows and light. Savannas, on the other hand, are both wet and dry in seasonal turn, and their plants are even-tempered too.

A cactus garden in Mexico

Conditions there are the result of two dominant forces that alternate in power: wet and dry tropical air masses that create annual rainfall of less than 1 inch in every month. Finally, savannas encircle the tropics in the latitude range farther from the equator than the rainforests, about 20 degrees north and south. They are found from the northern coast of Australia northward to India and Indochina, and from South America to Africa. The third tropical biome is dry and arid. This is the 12 percent of Earth's surface called "tropical deserts." These low-elevation areas are focused around the Tropic of Cancer and the Tropic of Capricorn, north and south of the equator between 18 and 28 degrees. The trade winds there blow earthward and dry the air relentlessly. The deserts include areas of Mexico and the southwestern United States, Argentina, Africa, and central Australia, where the native plants are as tough as the weather conditions.

Know and Grow

By understanding a bit about the native habitats that gave rise to the plants featured in this book, you gain insight into their growing needs and to your task as their steward. Essentially, tropical plants represent the four essential native realms: sunny and wet, sunny and dry, shady and wet, and shady and dry. Think of these designations as points along a spectrum of growing conditions with gradations in between: sunny, partly sunny, partly shady, and shady. Not to press the point, since it might seem obvious, but plant labels and descriptions can be rather vague on this subject. Full sun is six or more hours daily; less than six hours is considered partly sunny. This is sunnier than partly shady, which has brighter light than shady, but still no direct sun. When it comes to the subtleties of wet and dry conditions, the lines are clearer, at least for the plants in this book. They come from soil conditions and watering practices that stay consistently wet, dry out slightly, or dry out thoroughly between irrigation cycles. A few can become drought tolerant in maturity, and a very few can be grown in water, but they are also well suited to less extreme garden sites. Knowing a bit about a plant's origins is good for conversation, but, more important, this knowledge can lead directly to more growing success.

Many tropical plants will thrive in environments unlike their own, and gardeners take advantage of that every day, growing tropicals as annuals and potted plants where they are not winter hardy.

Even a climate-controlled office only slightly resembles a low-light, dry-side native environment, yet plants grow on. While they may be smaller versions of themselves, they retain the drama of the full-sized specimen. Some plants are rigid in their needs, but most can adapt at least a step or two across the spectrum. For example, a plant may be a compact bush covered in flowers when grown in full sun with plenty of water, but it can be perfectly happy with slightly less of each. The trade-off is a plant with a looser growth habit and almost as many flowers that needs less water to maintain its good looks. Likewise, a shade-loving, dry-side plant can often handle some sunlight, especially in more northern latitudes; it may even need some sun in order to stay dry and warm where summers are more moderate than its native environs.

Classic Favorites

Classic tropical icons such as plumeria, hibiscus, bird of paradise, and orchids are so recognizable that even children can draw their silhouettes. They are familiar symbols of all that is tropical, a status achieved through universal distribution of their flowers in fragrant leis and fabulous floral arrangements. Yet the tropical plants gathered here go far beyond those flamboyant flowers to intricate vines and chunky trunks. These diverse plants are united by the way they demonstrate tropical style, defined in their flowers and leaves. Tropical flowers, even the daisies, are over-the-top beautiful to our eyes, but their grandeur is born of necessity. They evolve to survive the rigors

of the tropical climate, to deter predators, and to attract suitable pollinators. Best described as complicated and sumptuous, the flowers are often waxy or silky and have rich, saturated colors that can withstand extremes of sun, shade, and heat. Some are oddly fragile to look at, yet they have incredible tensile strength and can stand up to the meanest thunderstorms.

Bird of Paradise (Strelitzia reginae)

More to Love

Large and small tropical flowers are noted for their long bloom seasons and superior vase life; bold colors dominate the palette, but even pastels hold their striking colors for weeks if not months. The combination of these qualities in one tropical flower surprises with its constant feast for the eyes, but many blooms go further to elate and seduce with fragrance. Gardenias, true jasmines, frangipani, and citruses grab you with aromas you could identify with your eyes closed. People and perfumers treasure the sweet, spicy, blissfully floral, and cleanly bright odors of those classic flowers. More scents abound in the tropics, from the deeply spiced cinnamon tree to the intense nighttime fragrances of night-blooming jasmine and moonflower. Nature's perfumes delivered on tropical breezes are transforming in very real ways, as evidenced by their popularity in toiletries and aromatherapy. These flowers beg to be carried and worn, in bouquets, in décolletage, and in leis. Industries are built in efforts to bring their fragrances to the world. Perfume designers have moments of pure inspiration as they formulate scents from flowers, and colleges of chemists spend entire careers trying to reproduce them.

As intoxicating as the fragrances of tropical flowers can be, these aromas are actually integral to the survival of their plant. Flowers have aromas because they have unstable, or volatile, chemicals in them that evaporate under the right conditions. These chemicals are triggered by heat, humidity, sunlight, or lack of sunlight and work like neon signs over a beer joint to welcome patrons. In the case of flowers, scent deters critters that might eat them and informs welcome pollinators that nectar and pollen are nearby. We humans consider ourselves lucky to be equipped with olfactory organs that enable us to enjoy them as well. But our attraction to them acts as another method to ensure their survival, doesn't it? It is hard to imagine a world where these fragrant flowers were allowed to become extinct.

As with the blossoms, the leaves of tropical plants display essential traits that define their style. They are almost universally big—at least in proportion to the plant they cover—in order to enable efficient water management in extreme conditions. Large leaves dampen the effects of flood *and* drought, particularly when they are thick and fleshy. They can grow into dense masses that provide shade and shelter underneath and within them. Evergreen tropical leaves often shine like the sun

they reflect and offer elegant lines that demand attention when flowers are not present. When not sleek, leaves can be deliciously hairy or madly crinkled, netted, and deeply veined. These last qualities combine into waffled leaves that cleverly shed and trap water simultaneously. Other tropical leaves are cut into chunky hands, long fingered leaflets, and clubby fist shapes that rustle and wave to entice passersby of many species. A select group of leaves use color to startle and attract, with deep, rich tones drawn into stained glass jewels and patterns so intricate they reflect the hand of a master painter. Some of these leaf

Flowers send signals to attract pollinators, such as butterflies.

Calathea (Calathea roseo-picta)

colorations are chevrons with feathered edges that decorate the whole surface, while other patterns envelop the leaf entirely in abstract arrangements. These leaves deserve their reputation for being "large and in charge" and personify the outrageous magnificence that is tropical plants.

Diverse Appeals

Another element of the tropical style is their mass, which gives them bold shape and establishes their visually robust form. Most of the plants included in this book are comfortably plump, dense with foliage and sturdy on their roots even when pruned or grown in containers where they are not hardy. Some are as well suited to bonsai culture as they are to a hotel atrium. To be sure, there are angry spines and mad thorns, gnarled trunks and unearthly forms in this garden too, each growing to guard its precious cargo or mask that treasure from view with an exotic appeal of its own. Not all of these plants are huge and stately, but the diminutive specimens prove that the loud tropical tone comes in small packages too, ready to be embraced by gardeners everywhere.

The qualities discussed here are not the only ones seen in plants native to the tropics, but they set the standard for interpreting the tropical style in gardens everywhere. Tropicals cannot stand temperatures much below freezing without damage, although quite a few can return from their roots in the subtropics, and a handful are hardy even farther north. Thus, while the plants profiled in this book do not represent every microclimate in the tropics, they made the list because of their high style points, ease of growth, and wide availability to home gardeners.

In Your Garden
Where and How Tropicals Grow

In addition to the tropical plants profiled in this book, there are hundreds more to tempt your garden taste. With the information provided here and your own good instincts, the only mystery about tropical plants will be how you gardened so long without more of them! For gardeners new to

tropicals, the photos will aid in identifying what's in the yard and lead you to the endless combinations possible there. Subtropical gardeners need only to know about them to want more tropical plants, and they will find helpful strategies for expanding their collections. As more tropicals become commonplace in the temperate zones, gardeners want a guidebook that demystifies their use there. Outdoors and inside, these tropical plants can make themselves right at home—wherever home may be.

Temperate Zone

Called the temperate or tepid zone, this part of Earth receives sunlight at an angle, so its warming effects are tempered. The zone technically exists north and south of the Tropics of Cancer and Capricorn, but to gardeners, the temperate zone excludes the border regions known as the subtropics and, more important, it is a climate unto itself. The temperate zone extends across vast areas of the world's oceans in the Southern Hemisphere, catching much of Australia, southern Africa, and a chunk of South America in its net. The northern temperate zone geography encompasses such varied areas as Alaska and France, Oklahoma City and

Kabul. Clearly microclimate variety is a hallmark of this zone, where winter low temperatures can average 32 degrees F or below 0 degrees F.

This is the land of four glorious seasons, where you can count on the spring thaw and schedule bus trips to see autumn's colors as the trees go dormant. A majority of the world's population lives in the temperate zones, mostly in the Northern Hemisphere, in part because conditions are not as extreme as in tropical or arctic zones. Summertime highs average in the 80s F, but the climates span a wider range than that number indicates. The definition applies to temperate climates wherever they occur, particularly at higher elevations in other zones. Freezing weather is expected and consistent here, allowing many plants and pests to go dormant for months. This environment is too cold for tropicals to survive without protection, even at their roots. There are many tropical plants that can grow outdoors in warm weather in the temperate climates and which then can be dug up and potted or propagated to move indoors when the weather turns cold. Many more tropicals are excellent container plants that easily move with the seasons. Look to the profiles for specifics about how to handle the plants in this collection during winter in the temperate zone.

Subtropical Zone

The subtropical zones comprise two narrow bands between the temperate and tropical zones. This region hugs the northern and southern borders of the tropics and runs along the Tropics of Cancer

Many tropical plants, such as these orange trees, can be grown in containers.

Tropical mandevilla vine (Mandevilla) *can easily be grown as an annual, no matter where you live.*

and Capricorn, to 40 degrees north and south, respectively. The subtropical climate is influenced by the weather generated along the equator to a much greater extent than is the temperate zone, as shown by its higher average temperature and rainfall. Climates here vary from deserts and savannas to monsoons and humid forests, but all are wetter than comparable climates in the temperate zones, especially in winter. Heat and humidity levels almost rival the tropics during the summer, which characteristically begins sooner and ends later than in the temperate zones. Plants native to the tropics are well known here because some are hardy, others return as perennials, and many are grown as annuals. The average temperature in the cold months is about 40 degrees, since freezes are seldom prolonged and sunny December days can be warm enough for shorts and flip-flops. The coldest weather usually begins at the first of the year and ends before March arrives. Lean-to greenhouses attached to homes are common here, since a window that opens onto a heated room is usually enough to protect any tender specimens in the colder months.

There are not four defined seasons in the subtropical zones, and drastic changes in temperature are common at the edges of winter. In North America, the subtropical regions encompass most of Florida, the coastal areas along the Gulf of Mexico, the deserts of the southwestern United States at low elevations, and northern Mexico. Subtropical climates include far-flung Northern Hemisphere locations from Savannah, Georgia, to New Delhi, India, and to Hanoi, Vietnam. Step off a plane south of the equator in cities from Sao Paolo, Brazil, to New South Wales, Australia, and smile— you are safely in the warmth of the subtropics.

Tropical Zones

Much about the tropical zone is covered elsewhere in this book, including its geographic location, zones, and microclimates. This region spreads north and south of the Equator and is also known, quite rightly, as the torrid zone. This is where Sol shines from directly overhead at least once every year. Locations range from the heights of Mount Kilimanjaro to the Atacama Desert, but the heart of the tropics lies in between.

The region is not all steel drums and boat drinks, but these icons add to the allure that attracts expatriates from every continent. As technology has evolved, more businesses can relocate to paradise and stay connected to offices anywhere else. With a higher birth rate and an overall trend toward inward migration, the tropics will dominate the world in population by the middle of the current century. Thousands of plant and animal species are native here, with new discoveries constantly adding to their ranks. Each time a new flora is identified, it holds unknown promise for medical research as well as gardening, which excites us all.

The plants known as tropicals are as varied as the places they come from, but the ones from lower elevations are not hardy in the temperate zone and many of these are iffy in the subtropics. The plants included in this book are a versatile and tough bunch, able to stand relentless heat and yet adapt to container growing for the indoor garden as easily as they do in their native soils. Perhaps this is because the tropics—wet or dry—do not have seasons in the way that northern latitudes traditionally do. Instead tropical plants have their own annual timetables that, depending on biome, force each one to develop survival strategies that translate to a variety of growing environments. The breathtaking results fool predators and welcome pollinators with flowers and foliage unlike any other on Earth.

The same tropical plant can go by several names, depending on how it is grown in a locality. It can be an annual—a plant that grows, blooms, and sets seed in one season. Bat face cuphea and candle bush tree are annuals by this definition, and they are allowed to naturally turn into compost when they are done in by frost outside the tropics. Luckily, they can be propagated or their seeds saved from year to year. A plant can be a perennial in the subtropics, like hidden ginger and night-blooming jasmine, reliably coming back from its roots each year, but be an annual in the temperate climes. A returning tropical is another term for a plant that can become perennial farther north, either from its roots or as a reseeder, like summer snapdragon and wax begonia. The names on plant labels can be confusing, but once you start growing, the basic principles apply to them all.

Growing Smarts: Starting Below Ground Level

One aspect of growing tropical plants that does not change with zones or weather is the soil. While the tropical plants in this book hail from around the world and from three main kinds of climates, they share the ability to grow in well-drained, fertile, organic soil whether it is provided in a container or a garden bed. The plant profiles tell you when a plant's care emphasizes one or another of these qualities and explain how to manage water and fertilizer to suit each plant. How is a nearly universal soil possibly suitable for such diverse plants that come from such different soils? Because it has what they need to grow and can be managed easily by the gardener with a simple maintenance routine. Furthermore, well-drained, fertile, organic soil nourishes the soil food web, which is the best-kept secret in gardening.

The Soil Food Web: Thanks, Dr. Ingham

With the advent of widespread use of electron microscopes, scientists are able to observe life at its most basic and revealing level. In work at the Oregon State University in the mid-1980s, Dr. Elaine Ingham delved into soil life at the microscopic level and began to unwrap the soil food web. A decade later, she started a business focused on the soil food webs of edible crops and the threats to them. The average shovelful of garden soil can appear to have fifty earthworms and more rolypolies, but at microscopic level, populations run into the millions. Nematodes, fungi, bacteria, and protozoa live on carbon in the soil; when plant roots are present, everything changes to create the soil food web. Roots exude carbohydrates that bacteria and fungi eat. They, in turn, are eaten by the "bigger" microscopics, protozoans, and worms that excrete what they cannot digest. In a perfect circle of life, roots absorb the excretions and grow, producing more carbohydrates to continue the cycle. This is the world of the rhizosphere, long known as a theory and now confirmed by the existence of the soil food web. The rhizosphere is a delicate place that feels the impact of every chemical introduced to it.

Healthy soil is more than a collection of sand, silt, clay, and organic matters. It's teeming with fungi, nematodes, bacteria, and protozoa, all part of the soil food web.

Enhancing Your Soil

Every garden soil can be amended to enhance the soil food web and, at the same time, create better conditions for tropical plants. The best organic potting mixes do a good job of encouraging the soil food web, and other commercial mixes can be amended with organic matters to improve them.

Even good potting mixes can be improved by adding organic matters.

Each plant profile has information about this process, but generally speaking, a good mix is three parts potting mix to two parts organic matters. Use a mix of different-sized organic matters, such as compost and ground bark, to improve soil structure at the same time. Even if you are not an organic gardener, your awareness of the rhizosphere and the fundamental need to nurture it should shape your garden choices. The incorporation of rotted mulch into the soil helps, and it provides some fertilizer as well. In addition, you need a complete, general-purpose fertilizer, one with major, minor, and trace elements made for use on garden plants. There are plenty of specialty formulas for various plant groups, such as palms and citruses, that can be useful, but there is seldom a need for much fertilizer variety. Many localities have particular favorite fertilizer blends, especially organic products. Since these are formulated with local conditions in mind, it is wise to pursue them. It makes sense to use organic fertilizers and pesticides because they intrude much less on the rhizosphere, but those who use synthetic products can compensate to an extent by replenishing organic matters regularly.

Microclimates: From a Plant's Eye View

Lots of tropical plants arrive as gifts and mementos of sunny sojourns with no concern as to who will grow them. Some folks will torture the green right out of them, but others find a new vocation through trial and error. The process for all of us comes down to figuring out the various microclimates in your space, whether we garden on an estate or a balcony. A simple adaptation of a landscape architect's concept of site survey can get you started in the right direction. The idea is to see the place from the plant's point of view, to see what's there relative to what you want. Artsy types can do it with a piece of plain white paper and a pencil, but the rest of us need a piece of graph paper, a pencil with an eraser, and a ruler or straight edge. Fold the paper into quarters and imagine you are above your property, looking down. Use the quadrants to orient yourself mentally. First, draw in the house, other structures, and anything else that is permanent, including hardscape elements like paths, pergolas, and benches, at approximately their scaled size and spacing. Add the neighbor's garage at the edge if it shades part of your garden. For indoor gardens, sketch in the walls, windows, and immovable furniture. Note where the sun comes up and where it sets and how water and wind flow across the place to pinpoint the most exposed and sheltered areas. Draw in water faucets, irrigation lines, outdoor lighting, existing beds, and evergreen trees and shrubs. Be sure to include the jungle gym, picnic table, and other moveable but essential accessories. At this point you have a good inventory of the site and can begin to assess and plan changes or simply celebrate what works. To learn more about conditions across the area, monitor temperature and humidity in several locations. Outdoors, set up a rain gauge to know how much water is available, natural or provided. Test the soil and the water (if it is known to be "hard" or especially rich in minerals) in new gardens by contacting your local Cooperative Extension Service for a kit. Step back and take a look at your drawing; all these steps come together to give you the knowledge you need to make more of your garden.

Alternative Strategies: Winter and Year-Round Tropical Growing

Any new plant can be a challenge to take on, but tropical plants present some issues that are predictable and readily resolved with alternative growing methods and strategies. Winter-specific solutions are included with the plant profiles and in the section on container gardening, but start with these:

Too many plants? Indoors or outside, grow upward. Shelving made of slats or pieces of wood, not solid wood, is called staging when it is used for plants, and it belongs in every indoor garden and most outdoor ones. These shelves make the most of the light available, but they also elevate pots for a better air circulation around each one. Gardeners often need room for another vine, and one trellis with three panels in a triangle shape usually fits where one panel did before. If even more growing space is needed, attach upright trellis panels to any fence to create vertical growing areas.

Chilly soil in spring? Temperate zone gardeners can see their warmest season shortened by a cool spring, and soils may stay too cool for tropical roots. Instead of waiting to plant while the clock ticks on your tropical garden, grow ahead of the season by using containers. As soon as the weather warms, nestle them into a bed of shredded bark mulch. Mulch insulates just slightly and may extend the outdoor season on the other end of summer as well.

Plants growing too fast? The simple answer is to prune and propagate. More information is available in the profiles for individual plants.

Sunlight baking your plants? When no shade is available to offer respite, install a simple structure and cover it with lath strips, trellis panels, or shade cloth. Light shade can be achieved with lath running in one direction only, 1-inch-wide lath spaced 2 inches apart. Trellis panels double the amount of shade your structure can provide. Shade cloth is woven fabric found in 20 percent shade and thicker gradations. It is less durable than lath or trellising but is excellent for temporary shade, such as an area to root cuttings.

Too little sun? This is easily remedied indoors with supplemental lighting. And while adding sun may seem impossible outdoors, it is not. Prune overhead branches to thin a canopy or remove one of every three shrubs in a hedge to let in more light. Change the orientation of a trellis so the sun moves over it instead of hitting it head-on to make room for more plants where there used to be shade.

Pushing the Envelope on Zones

Adapting your space to accommodate the needs of plants goes hand in glove with learning more about what the plants can tolerate. Often that lesson includes stretching the conditions a plant can take; as long as the adjustments are within tolerance, it works.

Still, trying to grow plants and flowers outside their comfort zone intrigues many people. We do it when we force paperwhite narcissus bulbs to bloom in a bowl of rocks as surely as when we coax staghorn ferns to grow inside wire cages affixed to boards. When it comes to tropical plants, gardeners who live where the winters are mild may take a chance and leave marginally hardy plants in the ground or in their pots outside. The plant may not die down completely, and if you have rooted a cutting or otherwise propagated it, there will be no great loss if an occasional bad winter kills the plant left outside. Plants hardy into the subtropics can be treated this way even farther north. These plants at a hardiness zone's edge and tender collections in large pots sometimes find themselves protected under a makeshift plastic cover or happily live within walled courtyards.

Try it; stretch your plant's wings a bit. In the changing climate of the twenty-first century, the dates of the first and last frosts have been observed to be shifting by a few days and summer heat records are broken regularly. These changes are confirmed in the 2012 USDA Hardiness Zone map, which graphically illustrates the expansion northward of the tropical and subtropical zones in the United States since the previous 1990 map. See the Resources section for more information about this map. More or less rainfall, extreme storms, and unusually windy conditions make the news weekly to drive home the point that conditions are volatile. We are challenged to find plants that suit these warmer conditions, and tropicals do not disappoint. It may be that your hanging basket fern can stay on the porch a week or two longer, or that Indian summer lingers to give candle bush plant a longer bloom outdoors in the fall. In the tropics, climate change can mean drier conditions and perhaps a need to add a lath structure to provide shade for parched plants that were comfortable in the landscape until recently.

Tropical Plants Are Part of Our Everyday Life

No matter where you live, it is quite likely that you will encounter an important crop from the tropics today. Ornamental ones are included in this book, but there are many more tropicals to know and enjoy. Citrus, sugar cane, and cotton are widely grown and even more widely distributed. The white mulberry tree, *Morus alba*, is cultivated for its leaves, choice food for commercial silkworm production.

But none of this group is exclusive to the tropics, as are important crops of banana, coconut, mango, and guava. Bananas and plantains, their firmer relatives, are part of the same *Musa* genus. Edible selections offer some range of colors and textures, but by far the grocery store banana gets the most attention and acreage. It is grown on over 10 million acres in more than 125 countries. You may eat a banana daily for its potassium content, or enjoy it as dessert, but coconut is an ingredient in everything from granola and Thai cuisine to toiletries and candles. Most of us take coconut for granted, but those who do not like its taste or fragrance work hard to avoid it. The popular coconut, harvested from palm trees in more than 90 nations, takes lots of space to grow. More than 26 million acres in the tropics are devoted to growing coconuts. Hugely popular with people who enjoy bright, tropical tastes, mango and guava are gaining in acreage annually. They are widely grown in the tropics and they are grown commercially for processing and export in south Florida, Mexico, and Asia. They are part of the burgeoning world of tropical horticulture which will only expand its reach as populations increase and the influence of tropical cultures spreads wider.

Agricultural crops receive that designation because they are grown on large acreages and are most often a monocrop. Huge farms or plantations in the tropics export the harvests of more than 5 million acres of tea, almost 20 million acres of cocoa, and a whopping 30 million acres of coffee to a caffeine-crazy world. Other important crops include yams and kidney beans, two that create confusion for gardeners and diners. Yams are cultivated in West Africa, Asia, and Latin America as a perennial crop whereas sweet potatoes are grown in the subtropics as an annual plant. Yams can weigh several pounds, are very starchy, and are a staple in the tropics seldom exported to the temperate zone. Kidney beans are smaller than red beans and are sometimes called chili beans. They are a dark red shell bean essential to cuisines as diverse as those of the Caribbean and northern India. If you have ever enjoyed red beans and rice on Monday in New Orleans, you know these beans; if you haven't, seek them out and celebrate the tropics any day of the week.

Controlling Pests and Diseases
Tips and Techniques

Besides the plants in your garden, there will be a wealth of critters—some to cherish and others to avoid if at all possible. Observing pests and diseases is one of the most basic reasons gardeners are

advised to take a daily walk through the garden, whether it is indoors or outside. You will be sure to see and smell your new flowers, but you will also notice changes in plants that could spell trouble if they're left unattended. Insect pests and diseases are not new to anyone's garden, and they are not especially attracted to tropical plants, but they *will* take advantage of their presence. Any particular threats to individual plants are included in their profiles so you can watch for them and take action as needed.

Generally speaking, there are two kinds of insect pests: those that pierce into and suck out plant juices, and those that bite off and chew plant parts. The first group can include aphids, mealy bugs, white flies, and the peskiest arachnid out there: spider mites. Spider mites are small and clever at masking their presence on plants until their damage is well under way. In small numbers they are no real threat, but most multiply exponentially at eight-day intervals to build *huge* populations in a hurry. The chewing pests are primarily caterpillar larvae and assorted beetles, but these can include bigger mouths like grasshoppers. Some eat randomly from plant to plant and are seldom a death threat, but others have

27

a preferred host and will devour it in a few days' time if left unchecked. Unless a plant is grown specifically as a host plant, such as milkweed for Monarch butterflies, it is up to the gardener to decide whether to let insect larvae feed or suppress them.

Trouble and Hope

It is good to know what an insect or other pest looks like, but given their propensity to run and hide, gardeners should also know the signs of their damage. Chewing insects, as well as slugs and snails, do visible damage in short order. Physical control by removing them is the first line of defense, followed by the introduction of predatory *Bacillus*, commonly called Bt, a barrier product to beat the slugs. Tiny insects and mites do more subtle damage and can be harder to spot, much less identify and control. Grab a hand-held magnifying glass to take a look at these pests, which seem to like the taste of tropicals:

Aphids, aka plant lice. Tiny as the head of a pin, round or pear-shaped, aphids are usually found in clusters at the growing points of plants and their flowers. They work hard to dehydrate plants, so leaves may be deformed, flowers won't open, and eventually growth stops. Fortunately aphids are slow and can be deterred by a blast of water from a garden hose. Larger infestations can be controlled by three applications of pyrethrin sprayed eight days apart. Before you spray *anything*, look closely to see if the aphids look puffy and gold, indicating that their predators are at work; their populations are being controlled and you should leave well enough alone unless your plants are suffering. When the balance between these competitors is in equilibrium, any sprays will upset it badly.

Scale and sooty mold

Mealybug infestation

Mealybugs. Sticky, white, living "cotton balls" form in a plant's crotches, where stems meet and leaves emerge, when mealybugs strike. If you go away for a month, a foot-tall green plant will become slick with the masses and bound for the trash bin. Luckily, mealybugs are easy to stop if you act early to paint them with rubbing or other alcohol. The cotton nest will then turn brown and can be removed with a paper towel. Afterward, lightly paint the affected area again, along with the nearby stems, to deter their return.

Whiteflies. Brush a whitefly-infected plant as you walk by and tiny white insects will fly quickly away. If you see one, you can be sure that there are scores flitting about elsewhere, ready to multiply. Tiny whitefly nymphs do not fly but hide, appearing as white specks on the undersides of leaves. When they feed in large numbers, the affected leaf will become pale and eventually turn yellow and drop. Watch for them on susceptible plants and get them while they can be got—i.e., before they are airborne. At that stage, three sprays of insecticidal soap with pyrethrin or an equivalent insecticide made eight days apart will control them. One evening at dusk, try to locate their main nest, usually located in an overgrown shrub. Disturb the nest and bugs regularly to further diminish their population.

Scale insects. This is such a varied group that can do so much damage! Cottony or slick, in a rainbow of colors, adult scale insects congregate wherever they can hide. They hide in leaf axils and in crevices on woody stems to protect themselves from predators, such as you and me. By the time you see this mess, the plant has likely dropped leaves. Six months before that point, the scale insects were little green nymphs, hatchlings from the lumpy adults. When you see them scurrying around the plant, they are easy to control with oil sprays.

Powdery mildew

Spider mites. Nearly microscopic but potentially perilous in hot, dry weather, spider mites result in crispy leaves that can take on bronze hues before they dehydrate completely. Infected plants will stop blooming and eventually fine webbing will appear in the leaf axils. Pruning and good sanitation are the ultimate controls, in conjunction with a pesticide containing clarified oil of Neem. Neem is an organic control useful for serious infestations of insects as well as mites, and it has some fungicide properties as well.

Diseases

Symptoms of common fungus diseases include these ugly events:

Sooty mold. A no more descriptive name was ever given to a fungus. Leaves and stems get a coating of smudge that slips off when rubbed. The mold is easily removed with soapy water, but you should look upward to find the true trouble. Feeding insects do not digest all they take in, and what's left creates a sugary surface perfect for the mold spores to land in and grow. Find the insects and control them to get rid of sooty mold.

Rust disease and leaf spot. When leaves develop spots, these can soon spread into brown patches as leaf spot diseases take over. Rust makes orange dots and streaks on the undersides of the leaves that should not be confused with other rust-colored plant structures like fern spores. See "What to Do" later in this chapter for control measures.

Root rot. Plants most often develop these diseases because they are overwatered. Leaves become pale and may turn yellow and drop off, often in the center of the plant first. Or they can turn brown on their tips and die back toward their stems. Fungicide sprays are less effective than drenches to slow the progress of these incurable diseases.

Pythium. This is the gray fuzzy stuff that forms on old flowers and stems left on a plant in cool, wet weather. Physical control by removing the ruined stems is all that is necessary to cure the disease, then avoid the conditions that cause it in the future. Pythium is one of the diseases that cause a condition called "damping off," the complete collapse of seedlings. Grow seedlings in a sterile seed starting mix, maintain good air circulation around pots and flats, water from the bottom, and do not overwater to prevent this disease.

Keep These Around

Beneficial critters should not be confused with their malicious counterparts. Watch what an animal is doing before you assume it is eating your plants, then take appropriate action. It seems counterintuitive, but there are fewer pollinators in tropical microclimates in proportion to the number of plants than there are in the temperate zone. Perhaps that explains why tropicals evolved with bloom and foliage colors like Las Vegas marquees. Surely that is how a tropical plant looks to the lowly ladybird beetle in your backyard. Look for these "good guys" and discover other local favorites; encourage them with good gardening practices:

Ladybug

Ladybugs. As seen in story and song, the adult ladybird beetle is a red-orange, round, hardback beetle with lots of spots. Efficient at eating other insects, primarily aphids, and multiplying to increase their fold, a colony of ladybugs is a godsend to a gardener. An invasion of them is difficult to deal with, however, and may require a vacuum cleaner to control if they get into your house. As with most families, not every relative is a good bug. Mexican bean beetles are copper or bronze and cucumber beetles have three rows of spots on their backs, and both will eat your plants.

True bugs. This trio can be scary to encounter in the garden until you see them at work. Larger than most, they are incredibly mobile, brandishing their swords. Their names reflect our fascination with them. Assassin bugs are recognizable by their red tummies, and damsel bugs have pinchers for front legs. These two prefer soft-bodied bugs for eating, while the beautiful, diamond-patterned pirate bug uses its long beak to devour thrips. Noble actors all, and worth cultivating.

Braconid wasps. Smile when you see little white spikes sticking to chewing caterpillars such as leaf rollers. These are braconid wasps, chomping away.

Praying mantid

Tachinid and syrphid flies. The little helicopters of the garden, the hoverflies, will consume some of the garden's worst pests, including the tachinid's favorite, stinkbugs. Tachinid flies are also known as sweat bees and can resemble that group, while syrphids are more wasplike but much smaller. Fly babies are called maggots; your opinion of that word may change when you learn that they eat aphids for lunch.

Praying mantids. A few are good for nibbling on aphids, but these hinged insects tend to be indiscriminate in their diet. They should be avoided in large numbers.

Lacewings. The graceful green adults with webbed wings are not much help, but their larvae are called "aphid lions" due to their appetite for aphids. It helps that they look mean, so you are less tempted to grab them in haste. Let them eat!

What to Do

Too often the issue of pest control comes down to spraying pesticides, or dumping the plant, or both in sequence. Those are, in fact, the last resorts and while they do belong in a gardener's arsenal, other strategies should precede them. The best way to manage the pest population in any garden over the long-term is to grow healthy plants in ways that sustain wildlife, including beneficial insects. That means gardening for the basic principles of backyard habitat: ready access to food and water, safe places to rest, and places to nest. By growing a variety of tropical plants, you provide these essentials for many critters, including any bee, dragonfly, lizard, or toad looking to pollinate flowers and consume insects. By varying the plants' heights and flower shapes and seriously limiting the use of pesticides, you'll attract the beneficial insects that can take on the common pests and keep their numbers in check.

The daily garden stroll is the finest short-term approach to pest control and can train your sharp eye so that issues never become problems. You will see that first sticky white mass of mealybugs on a stem or leaf axil or the first chewed leaf. At that point, physical methods (think "stomp and squish") and non-pesticide controls such as rubbing alcohol on a paintbrush or homemade pepper sprays work well. Larger infestations and elusive pests like whiteflies should be dealt with first by isolating the plant when possible and washing it completely with a mild solution of soapy water, administered by a dunking or with a cloth rag. If the pests persist, use a pesticide chosen for its ability to control *that* pest on *that* plant. Select a product that will be effective but not overly persistent and so will not pose a threat to beneficial insects that might drop by in the near future. The use of synthetic pesticides and fertilizers has implications for the soil food web too. The delicate balance between macro and micro life in the soil and plant roots deserves conservation.

Like insects, fungus diseases can attack both above- and belowground, but unlike bugs, control is not possible. Instead, diseases will be suppressed and their further development stifled using control methods such as pruning, sanitation, and fungicide sprays and dusts. Remove an affected leaf or flower as soon as you notice the invasion and watch the plant carefully. When plants are found to be susceptible to fungus attacks, keep their leaves as dry as possible and prune to thin their dense growth to improve air circulation to slow disease progress.

Fungus diseases can be controlled by using appropriate fungicide sprays.

Ask the Expert: Pest FAQs

Question: There are sticky white, cottony things on the stems of the tropical plants in our office atrium. What are they, and what can we do?

Answer: Mealybugs look just as you describe and populations can build quickly indoors. First, separate the plants with mealybugs from those that do not have them and place all of the pots so none is touching another. Use a cotton swab soaked in rubbing alcohol to "paint" all of the mealybugs until they turn brown and then remove their remains with paper towels. Spray those plants well with an insecticide labeled for mealybug control and for indoor use. Spray every 8 days for one month.

Question: I planted a bed of angel wing begonias and coleus and right away, something is eating the little plants. Can I save them?

Answer: Slugs and snails can quickly devastate tender young plants but usually begin feeding on the leaves. If the growing point at the center of your transplants is intact, they can recover with close attention from you for fertilizer and water needs. Get a slug control product that is not toxic to birds and spread it liberally in the newly planted area. Most important, locate their daytime hideouts. Follow their slime trails or look around for damp leaf piles, shady concrete benches, and cool spots under the porch. Rake out the leaves and use soapy water or sprinkle lime to make other nests less hospitable.

Question: The leaves of my orchid plant look almost sunburned and seem to have webs growing under them. Is there an insect to blame?

Answer: Spider mites spin webs on dry surfaces like the undersides of your orchid leaves after they cause other damage including leaf discoloration. These arachnids are almost too small to see until they move and will sometimes scurry out of the way when you water. In fact, regular watering and occasional misting under the leaves can help prevent spider mites. Spray a few times with Neem or an insecticidal soap. The schedule depends on your product, but you will need to follow the label directions closely to gain control over these pests.

Question: A black mold is growing on the elephant ear plants in my shady garden. What would cause this?

Answer: Sooty mold is the common name for the fungus that coats many leaves in summer. It is the result of insects feeding on the plant or on one above it. What aphids, whiteflies, scale, and other insects cannot digest, they excrete. The substance covers leaves and even lawn furniture, ready for sooty mold to find a new home in it. Clean it off with a spray of soapy water, find the insects and control them to send the mold packing.

Question: I brought a lot of plants in pots into my house and a few days later they were swarming with gnats. How can I get rid of them?

Answer: Fungus gnats hatch and swarm when conditions are damp and warm, such as well-watered container plants in a warm room. A drench of any contact insecticide suitable for use on your plants will control them. Mix it at regular strength and apply enough so that the solution runs out the pot's drain hole. Many gardeners do a preventative drench each fall before bringing pots into the indoor garden.

Question: Every day there are holes and torn edges on the leaves of my split leaf philodendron. What should I be looking for, and how do I treat it?

Answer: Caterpillars, the larvae of numerous butterflies and moths, are serious eaters and prefer green, watery tissue like your leaves. Turn over every leaf and inspect every stem to find them; stomp and squish the offenders and then introduce a natural control such as Dipel. Keep an eye out for black droppings, or frass, and for egg masses in symmetric clusters on the undersides of leaves.

Question: Recently some gray spots appeared on the staghorn fern that I hang in an oak tree. They seem to be limited to the new spathes, but how do I get rid of them?

Answer: Leaf spots can be the symptoms of fungus or bacterial diseases that can be hard to control. Remove any plant parts that are affected, and if there are multiple staghorn ferns growing together, separate them to improve air circulation around each one. If they get no good light at all, relocate the plants to a brighter location to allow them to dry out more rapidly when rained on or watered. It would be wise to spray and drench the staghorn ferns with a fungicide, but keep an eye on the plants.

Question: The flamingo flowers and gerber daisies that I planted can't seem to bloom even though the plants look fine. The buds are there, but then they don't open. What is the problem?

Answer: Two common insects can stop the final stage of flower development and prevent buds from opening properly or at all. Aphids are pinhead-sized bumps that you can see on the outside of the buds but thrips hide inside. Inspect the buds for aphids and cut some open to locate the thrips, which look like white commas inside the flower bud. Treat the plants for both insects with a fortified pyrethrin spray such as one that has added potassium salts of fatty acids, or with Neem oil. If the problem persists, consider using a systemic insecticide.

Propagation
The More the Merrier

Gardeners take to propagation like birds to an oscillating sprinkler on a hot day. We make more of our tropical plants for several reasons, beginning with the pure joy of doing so. Beyond that, it is a way to get more plants to fill a bed or additional pots. Propagation is certainly the wisest way to grow some tropicals, particularly vines, from season to season where they are tender. The smaller, younger plants can better adapt to indoor conditions. The plants in this book can be propagated by one or more of these methods: cutting, division, layering, and seeds.

Cutting. There are three kinds of woody cuttings commonly used to root tropical plants and two types of green-stem cuttings. All, except hardwood cuttings, can root in light, well-drained mix such as potting soil mixed equally with finely ground bark and perlite or coarse sand. Some, particularly green-stem cuttings, will root in water. Whether in a glass of water over the sink or in pots on the porch, herbaceous plants can be rooted from 4- to 6-inch-long tip cuttings and from stem cuttings about the same length, so long as there is a growing point, or node, included. Take the leaves off the lower half of a stem and plunge it into

damp rooting mix. If the leaves are large, cut each one in half, across its width, to reduce transpiration (the evaporation of water through its pores). A few plants included in this book have succulent stems that ooze sap when they're cut; unless otherwise indicated, these cuttings should be allowed to dry for a day before potting to root. They will also benefit from a sandier mix.

Woody cuttings are designated by their rigidity or lack of it: softwood is tender new growth, semi-hardwood is more rigid but will still bend, and hardwood will snap if you try to bend it. Softwood and semi-hardwood cuttings are most commonly used to root tropical plants. Ideally, the

Many green-stemmed plants can be propagated by division.

cuttings will be 4 to 6 inches long and taken from the tip. Strip the leaves off the lower half of the stem, cut its base on a slant to increase its surface area, and dust the cut surface with rooting hormone before sticking it into a small container of damp rooting mix. Keep the pots damp but not wet, out of direct sunlight, and top them with a plastic tent to increase humidity in dry climates. You can make a simple cloche (protective covering) for a 4-inch pot by recycling a 2-liter plastic bottle. Cut out its bottom but retain its cap. Slip this over the pot and remove the cap daily to allow it to ventilate and permit air exchange in the rooting chamber.

Division. Most green-stemmed, or herbaceous, plants can be multiplied by dividing their clump—the underground storage organs including tubers and rhizomes. Those clumps consist of top growth, roots, and the mass in the center that sprouts both—its crown. To propagate a new plant is a matter of separating the crown, retaining a bit of the top and roots with each division. The storage organs can be cut apart or separated, depending on the plant, and replanted immediately.

Layering, air and ground. Some tropical plants have reedy stems, grown from canes or soft-wood branches. They are good candidates for air layering, the time-honored practice of growing a new plant. It is particularly handy when a mature plant has leaves above and below an area of naked cane. Make a small slit in such a stem, wedge it open with a piece of toothpick, wrap it in a handful of wet, long-fiber sphagnum moss, and cover the whole business with clear plastic wrap anchored above and below the moss. When roots fill the plastic, cut the stem below the new roots to remove the new plant from its parent.

Ground layering works for plants with flexible stems near ground level. Dig a shallow trench in the ground near the plant's base and nestle a middle portion of the stem in it by bending the stem to form a croquet-type wicket or hoop. To one side of the trench is the mother plant; to the other is the "new" plant's leafy tip. Use a brick or board to hold the future roots in place if needed. Keep the layer watered and watch for new growth at the tip to signal that the process is complete. When that happens, the new plant can be cut away to grow on its own.

Seed. The entire future of a plant lies in its seed, and growing them is a noble vocation. It is also a fine way to share the wealth of your garden. Most of us are not Johnny Appleseed, able to just sow and grow, and we find more success by starting seedlings indoors for timely transplant to the garden. The percentages of healthy seedlings you can produce goes up dramatically with three simple additions to your growing regime: First, improve the growing conditions with a self-contained seed-starting setup, preferably one with a light fixture and a hood to increase humidity around the seedlings. Next, get some peat pots and sterile seed-starting mix or soilless starting mix with vermiculite or perlite and use this to give seeds the freshest, most disease-free start possible. Finally, discover bottom heat for faster sprouting and bottom watering for growing seedlings. An inexpensive electric mat made for this purpose, not a conventional heating pad, provides constant warmth to flats of seedlings. Once the seeds are up, put a tray under the seedlings to fill with water and allow the peat pots or pellets to take up what they need without dousing the babies and knocking them flat with overhead watering. Keep the light source 3 to 4 inches above growing plants.

Propagation Methods for Tropical Plants

The following are common propagation techniques for tropical plants, and the plants they work best for. Look for more propagation recommendations in the plant profiles.

Method	Use For
Stem Cuttings	cane and angelwing begonias, dumb cane, dragon tree, firecracker flower
Tip Cuttings	blunt leaf peperomia, wax plant, nerve plant, schefflera, scarlet sage
Offsets	staghorn fern, snake plant, giant aloe, pineapple, silver vase bromeliad
Divisions	Philippine orchid, gerber daisy, nun's orchid, bird of paradise, flamingo flower
Ground Layers	firecracker plant, shower orchid, bougainvillea, mandevilla, allamanda
Air Layers	weeping fig, Arabian jasmine, Chinese hibiscus, bottlebrush
Seeds	candlestick plant, hyacinth bean, queen of the night, moonflower, papaya

Other Plant Parts

Back bulbs and keiki	orchids
Rhizomes	Rex begonia, upright elephant ear
Tubers	caladium, clivia, elephant ear

Design and Garden Uses
Why We Love Tropicals

Most people remember one tropical flower that sparked a lifetime love affair with these blossoms. Yours might be the hibiscus in Esther Williams's hair in an old movie or the gardenia behind Lady Day's ear on an album cover. Some of us remember the orchid corsage for prom, a crown of gerber daisies on a young dancer, or a greenhouse wall covered in golden trumpet vine on a January morning. Brides never forget the aroma of orange blossoms or Madagascar jasmine (*Stephanotis*) and the perfect white flowers both bring to the bouquet. For military veterans, it might be the sweet aroma of a well-deserved leave, disguised as frangipani leis at the Honolulu airport. These memories have burned images into our souls and solidified our connection to them. We celebrate them because of their associations to our lives, but it is their sheer power as design elements that firmly establishes tropical flowers above all others in their ability to stir our emotions.

Fat, Fragrant Flowers

Tropical flowers grab our attention with bold colors that entice us to linger in their presence, pluck a bunch, and hold their fragrance close. One reason for this magnetism is often the sheer size of these flowers, either as huge individuals or lush clusters. You might think their girth is the result of year-round growing conditions or genetic freakism, but there is order in Nature. You can see the grand scheme displayed on tropical flowers with oversized petals or gigantic bracts with tiny flowers at their heart. The petals and bracts are there to protect the precious true flowers from predators and to attract pollinators. For example, in the Araceae family, tiny blooms are packed onto a long nose of a tall flower spike. They would be easy pickings for browsing wildlife if it were not for the bountiful bract around each bloom spike that offers shelter to pollinating beetles. Those colorful bracts are hard-working plant parts—modified leaves actually—and their variety is endlessly fascinating across plant families. Expansive petals have a similar role, acting as a flashing sign to those who should gain the nectar while thwarting unwanted visitors who might contaminate the all-important pollen. The petals' relative vastness also exposes enemies as they approach and leaves them open to a counterattack from beneficial insects. A flower's fragrance, ethereal as it may be to us, is sometimes a part of this repellant dynamic and apparently will hold off some potential predators while attracting pollinators.

Bells and Whistles

The large and small parts of tropical flowers have equal power to attract us and boggle our minds just looking at them. You can get stuck for days trying to figure out how flower petals get stacked up like they do, or why butterfly flowers have stamens that look like antennae. Tropical blooms are endlessly entertaining and make us wonder why their unthinkable complexity exists. The tubular flower types

Blue butterfly flower (Rotheca myricoides)

Hibiscus (Hibiscus rosa-sinensis)

range from tiny to tremendous and flare open like the scalloped horns in angel trumpet, or stay closed as in the circular clusters of lion's ear. Others tubes lead into flower chambers so exotic they make you want to shrink yourself to explore them firsthand. When botanists observe and dissect one of these specimens, such as a corkscrew flower, they reveal a natural maze that a potential pollinator must traverse only from the left side to ensure its cleanliness. As might be predicted, the daisy-type flowers are an elaborate take on that basic disk and ray flower arrangement. Gazania, for example, puts on layers of long and short multicolored ray petals that stick out like the brim of a sombrero around raised, fuzzy central disks. Their beauty and our curiosity make a good match, and our affection for them keeps us both intrigued and amused.

Long Seasons

Gardeners have two points of view about plants that bloom only once a year: we either wait with baited breath and call the neighbors when the night-blooming cereus goes off, or we dismiss the single roses in favor of the remontant—those that bloom repeatedly. A majority of tropical flowers bloom for weeks or even months in season, or off and on all year. This delectable habit results in a plant like fairy petticoat that may have flowers and fruit present together. It can mean flamingo flowers in the border all summer and through the winter, even where snow is on the ground outside. Tropical flowers are not among those that make us wait for years to see a bloom, either. We are enchanted by the long bloom seasons, early maturity, and the longevity of many individual flowers. These flower qualities function to increase the chances of pollination by extending the time that nectar and pollen are produced. It just so happens that this way of ensuring a fresh pantry for pollinators whenever they arrive also makes for some of the best choices for our gardens too.

Red salvia (Salvia coccinea *'Lady in Red'*)

Colorful Dimensions

Tropical plants are nothing if not hospitable, rolling out the welcome mat for the pollinators they hope to attract. Leaves and flowers boast a 64-color box of crayons sure to have a shade that calls the right tune—and we sing right along. No one knows if the nectar-lovers have emotion, but their favorite colors certainly play on ours. Every shade of green soothes and cools one's mood just as a crazy quilt of mixed colors sets a busy tone that stirs a need to act. In tropical plants, color adds another dimension altogether. Like three coats of polish on a full set of acrylic nails, the color in tropical plants looks thicker and often glossier than would seem natural. The added dimension of their high-gloss hues may have evolved along with their environment, but they provide design strengths every gardener can put to use.

Full Spectrum Appeal

The power of red to move people in one direction or another is well documented, and a row of scarlet sage proves it. We feel motivated to walk its length, no matter where it leads. If you do not want to call attention to your side gate, leave red out of that bed. Pure red is too hot to resist and is the color most apt to get a rise out of you. The farther red ranges on the spectrum towards blue, you find the excitable shades of purple passion from fragrant heliotrope to summer snapdragons. Blues—azure to navy—bring a sense of oceanic calm and inspire feelings of introspection. Sky flower's clear blue hue cools a hot patio immediately, and everybody feels better with it in sight. Emerald creeper's turquoise blue, though it begins to flirt with yellow, induces an equally cool attitude coupled with unbridled joy. Going in the other direction, where red heads toward yellow, are the colors of loyalty and steadfast friendship, personified by orange. There is a sense of certainty about a fence of Mexican flame vine when its bright orange flowers put on their show. Your emotions tell you all is right with the world. Yellow itself is the most optimistic of colors and engenders a sense of cleanliness and good cheer. That is why a front bed lined with yellow bells can make you feel good about a house whose interior you have not yet seen.

Surprisingly Bold

There are few pastels in the tropical color palette and its pinks hardly qualify either. They are rose-pink, mauve, salmon, and many more shades, some almost indescribably deep and moving. They are the sitars of the tropical garden, a raga to enchant and elate you. Surround a seating area with pink sultana and Persian shield to make visitors comfortable and perhaps facilitate fanciful conversation about something other than work. Even white can be a bold color when it is laid on thick, waxy flower petals like frangipani and queen of the night. The color white adds "starch" to tropical plants and the flowers stand up to the other hot, bright colors.

Deceiving Looks

Texture can be an elusive quality for those unfamiliar with its effects on a design, but its mysteries are simple to understand. Anyone who has picked out upholstered furniture or selected fabric from a shop full of cloth bolts knows about texture. It is the visual heft of any object and is a measure of its tactile impact on the viewer, not its actual components. In terms of fabric, wool tweed may look heavy and rough but actually be relatively thin and smooth as it flows off the bolt. It has a coarse texture based on its superficial features. The same principle applies in the garden, where texture is the apparent way a plant would feel if you ran your hand over it. It is the way you see that plant, not its true substance. For example, the long, thin, almost leafless stems of firecracker plant look like they would feel soft and seem to drape gracefully, like a fiber optic fountain. In reality these stems are stiff and gravity pulls them down to balance the weight of their red tubular flowers. Firecracker plant's texture is fine, because it looks that way, but the leaf scars along the stem will feel quite rough.

There are as many variations on texture as there are people to look at plants, but almost all of them fall into one of these general categories:

Fine texture bids you to come closer to examine its details. It looks like the plant would feel soft if you brushed by it, or as if it had been drawn by a fine-point pen. Besides the firecracker plant, fine texture is the province of false aralia and many ferns.

Coarse texture looks like it would hurt to touch it, as if it would be rough like many tree barks. Its effects are obvious at a distance, easy to see like cartoon characters with broad

The narrow leaves of firecracker plant (Russelia equisetiformis) *are striking.*

Tropical plant leaves have an enormous range of textures.

outlines. Besides tree bark, look for it in big leaves like plants in *Colocasia* genus, those shaped like blunt daggers, and flowers with exaggerated parts like Chinese violet.

Smooth texture can be found in leaves and large flower petals with few veins showing and sleek patches of color. Its visual weight can be as effervescent as a silk parachute or as dense as peanut butter, so long as it looks smooth and seamless. It is found in flamingo flower and cat's whiskers, among many other plants with slick leaves or flowers.

Rough texture gives the illusion of three dimensions, delightful tricks of the eye brought into being by the veins in a leaf or flower. Like the thread in a lumpy seam, each vein of plants like elephant ear seems sewn too tightly and draws up puckers and pleats across the surface. Add colors and hairs to this waffling effect, as Rex begonia does, to see the best in rough texture.

Functional Texture

Incorporating texture into a design can be as simple as planting Cape daisies in front of a Chinese violet on a trellis. All parts of the vine are fleshy, heart-shaped leaves and yellow flowers in clusters that hang down as if they were heavy. Its growth is dense and its texture is moderately coarse, as if it were drawn by a grease pencil. In comparison, the low-growing daisies are a symphony in fine texture. Narrow, upright, pointy leaves support their carpet of short-stemmed daisies, whose perky, pointed petals reinforce the lighthearted look. The contradiction in texture creates more interest in the scene, relaxes the vine's intense effect, and puts some muscle behind the daisies to be sure they are not overlooked.

Lines define space to our eyes, and they cause us to shift focus and change the way we feel. By using the lines of tropical plants to establish boundaries and embellish scenes, you can display a range of emotions worthy of the best novel. Lines move people in more ways than one. Water gardeners know that tall plants belong pondside to unite earth and sky in a symbolic way. It is their upright lines—the trunk of a palm or dragon tree—that literally lifts the eyes and so inspires loftier thoughts.

Converging lines, however, create confusion and may bring on the feeling of being a bit unsettled, ready to go without a destination to look toward. Like trying to read an unfamiliar map, the maze of lines that creates patterns on coleus and croton can confuse us, but can also be a call to action. Curved lines can be as regular as a scalloped collar or undulating, sexy, and provocative. The first sort can seem instantly familiar, a déjà vu moment even when the plant displays extremely dynamic lines, as on lobster claw flowers. Once you have seen one, you know what to expect in the rest, inspiring confidence. Longer, less predictable curved lines—like a hedge of calico plant or jacobinia—will slow your pace and yet propel you on to see what is around the next curve. Intrigue and seduction are the emotions carried into the garden by irregularly curved lines, while straight lines and right angles offer clarity and reliability. Like a close rank military drill, the lines of reed stem orchid and bottlebrush are predictable and comfortable. You know what to expect of straight lines and the plant stalwarts speed you along your way.

Concrete Forms

The design element called "form" is a way of applying solid geometry to the real world, and tropical plants are its perfect vehicle. Form provides distant views, as when a row of miniature date palms

The architectural forms of large tropical plants can create a unique focal point.

behind the pool becomes a wall of upright fountain shapes. A sturdy evergreen form like bush lily or dumb cane establishes an optimistic focal point with its widespread, welcoming shape. Some forms are natural, while others are the result of pruning or training a plant onto a structure of a particular shape. Triangular forms are the steeples of the garden, meant to be spiritually uplifting. Spineless yucca offers a dynamic triangular form, both overall and in individual leaves, and brings reverent silence with it. Full circles feel complete, embodied in pinwheel flowers like yellow gardenia and gazania. Square shapes are friendly and reliable, like armchairs, while rectangles are dicier and move us to extend our vision beyond them. The square stems of the blue butterfly flower and other mint family members certainly live up to their billing, as do the bulky, square forms staghorn ferns so often develop. Both are comforting plants because their form lends stability to any design. Large, rectangular shapes usually derive from pruning or structures and none has more power than a pergola covered in vines such as flowering pandanus and pink trumpet vine. The combination of rectangular form and pink color in one design begs for a hammock to further appreciate its laidback, confident emotion. Like planned spontaneity of some vacations, irregular shapes can be reliably unpredictable. For example, we know that queen of the night will have zigzag branches, but the form they ultimately assume is impossible to know until they have grown. The irregular forms stir emotions of wonder and puzzlement to keep us looking at their designs.

Combining Plants for Contrast and Complement

Contrast and complement go together in designs to achieve harmony from diversity and create more interesting scenes. When two plants contrast sharply and effectively, each calls attention to the other as the viewer makes unconscious mental comparisons between them. The sharp lines of sanchezia, for instance, will contrast with almost any other plant, but it takes a special plant, such as copperleaf, to hold the stage with it. While its color is less strident, the copperleaf's patterns and variegation hold their own with sanchezia: palm-shaped leaves versus ovals, overall abstract leaf design versus stripes, rounded plant shape versus angular form. Neither overshadows the other and each looks better for having a foil. Remember that plants within the same view of the gardener should have friendly contrast; avoid boxing matches between plants growing near one another.

Copperleaf (Acalypha wilkesiana)

The popular wax begonia shown bordering this water feature is a tropical plant.

Like complementary colors, plants can be opposite each other on the color wheel and somehow still go together. When that happens, it is usually because the shades between them—the tertiary colors on the wheel—are included in the scene. In the example described, sanchezia is a green-and-yellow-striped delight while copperleaf relies heavily on the red palette with tones from pink to gold. Sanchezia's green and copperleaf's red are opposite each other on the color wheel and no matter what time of year it is, together they evoke Christmas and not much else. Fill the huge gap between them with the yellow of sanchezia matched by the gold-orange copperleaf and suddenly they complement, or fulfill, each other.

Mix, Not Match

The art of combining plants for the best effect is the subject of whole books and occupies entire careers. It helps to think of plants as contrasting types that can be drawn into complementary designs. Some obvious groups include plants with:

Distinct forms and shapes. These are the iconic plants and other strong profiles that deliver tropical style by their very appearance in the garden. Consider the palm and bird of paradise, but also plants like papaya and bush lily to ground a design with their sturdy, sheerly tropical silhouettes.

Glory lily (Gloriosa superba)

Buckets of flowers. For natural contrast with crazily painted leaves, choose plants that bloom all over, in the leaf axils as well as at the tips of stems, such as impatiens and some jasmines. Favor those that form pendulous clusters, such as horn of plenty and Philippine orchid, and lots of upright bristles, such as golden brush ginger.

Fast-growing bold ones. One strength of tropical plants is their rapid rate of growth, a feature that can change the looks of a garden quickly. It is impossible to overlook an Abyssinian banana or an upright elephant ear, and nerve plant or Persian shield that will fill a space while your back is turned.

Little jewels. Dwarf varieties offer no less drama and can grow where their larger siblings could never fit. For example, the true date palm matures at 100 feet tall, while the miniature date palm tops out at 10 to 12 feet with a similar chunky trunk and graceful crown. Cane and rhizomatous begonias range widely in size, including diminutive varieties that ensure there is one for every garden.

Vertical works. Grow upward to gain strong lines and fabulous flower displays, but let some vines slide into ground cover status or spill from the top of a wall. Mexican flame vine and glory lily are especially versatile, as are hyacinth bean and moonflower vines.

You will no doubt think of other groups, like plants with leaflets that can lighten the overall texture of a composition or red-flowered plants, simply because you like them. Separating plants by their key feature keeps you from crowding one scene with all of one type, but also guides you to use different sorts for powerful contrast.

There is one more group to consider: the bullies and rampant reseeders. These are plants that are not technically invasive but which gardeners are often advised to avoid, such as Kashmir bouquet (*Clerodendrum bungei*) or Mexican hat (*Ratibida*). Neither is included in this book, but like many similar plants, they can be useful in areas where they have not been declared "invasive exotics." Grow them alone where the soil or drainage cannot be improved, or plant two in a big empty space and let them duke it out for the crown. Keep such plants contained and away from slower-growing desirable plants, and deadhead flowers before they can fling the next generation into your lawn.

Designs for Living

When it comes to selecting and placing tropical plants in garden vignettes, it is a wise idea to first determine the plant's purpose. The plants may have similar appeal or color or be the ones that you could not resist for a particular feature. Grow these last as a collection, such as begonias or plants

The bold, simple shapes of these leaves anchor these tropicals in this restful landscape.

The tropical landscape transforms every architecture into a warm, friendly scene.

with bracts. Groups of plants with patterned leaves do not usually make a pleasant design together because, like two plaids and a print in one outfit, they clash. Instead, spread that collection around to let each show its stripes or swirls to best advantage. One color can take charge of a garden room or mixed pot to dramatic effect, or be the signature color in a series of vignettes.

Outside of frost-free climates, tropical plants inspire thoughts of a summer oasis and combining them in a design accomplishes that. Use the tropicals as annuals in the same bed with those in containers with an eye toward the pots as hardscape and accents for the design. A bamboo palm potted in a squat red pot makes stunning centerpiece to a bed of caladiums. Their tubers can be lifted in the fall when the palm moves to a more protected location. By combining plants in pots with those in the bed, it is simple to adjust their maintenance routines if they have diverse needs.

Rule of Three

Just as the most stable figure in architecture is the triangle, outstanding designs rely heavily on the rule of three. You see them in floral arrangements, from the most minimalist representational piece to grand flower pyramids as big as a sink. A patch of moss, a clump of reeds, and one cluster of flowers is symbolic of spring if they're arranged the right way. To achieve the highest degrees of harmony in traditional Japanese floral design, there must be an upright form to dominate the view, a subordinate form to balance it, and a third form that blends the two together. In the same way, blunt-leaf peperomia can creep and spill while golden brush ginger towers above and nerve plant fills the space between them. While there are no hard and fast rules, the rule of three pleases the eye in landscape and container designs as well as in a floral arrangement.

Whether you are starting a new garden in the tropics, renovating an old one, or embellishing an established landscape, remember the rule of three. Keep to a straightforward plan that adds plants for a particular feature, like color or bloom time, and makes sense in the larger view.

Zone Smart

Gardeners in the subtropics can incorporate a tropical mix of returning perennials and annuals into their existing garden styles. Blending their attitudes and attributes in a cottage garden can mean extending its upright lines and offering strong forms to add to its stability. Red Abyssinian banana does both and echoes the popular cottage garden *Musa* species, while flaming glorybower has undeniable cottage charm. To blend tropicals into a more formal style, such as a Japanese-influenced garden, look to dwarf forms and relatively smaller woody plants such as allspice and cinnamon tree.

In the temperate zones, gardeners can add hardy, tropical-looking plants to sustain the mood while potted tropicals are safely indoors. The indoor garden can become its own paradise, particularly if you consider container gardening an integral part of interior design.

*Tropical plants included in this container arrangement include: hibiscus (*Hibiscus rosa-sinensis*), licorice plants (*Helichrysum petiolare*), verbenas (*Verbena 'Homestead Purple'*), and ivy geraniums (*Pelargonium 'Caliente Rose'*).*

Make the Most of Design

Make the most of your containers and garden beds by pairing colors and leaf shapes that create a mood that reflects your style. Bring together colors with similar intensities—bright with bright, for example—and use hues of the same color to draw attention up close. Besides color choices, leaf shapes work to establish your style through the garden whether leaves are solid green or wildly patterned. Like colors, you can use leaf shapes together or in endless combinations. Put the two together and patterned leaves can stand alone as focal points or be part of a rowdy band of tropical style. Like fashion, garden style depends on your good eye to edit the elements so each shows off the others to create a very personal and pleasing vision.

Color Effects

To reflect your comfortable, easy style, pair analogous colors. They are naturally cozy and sit next to each other on the color wheel.

Color Pairings	Example
orange and red	orange gerber daisy and red Abyssinian banana
blue and violet	skyflower and bamboo orchid
yellow and green	golden brush ginger and African mask

To feature contrast and spark conversation, mix complementary colors. They are opposite from each other on the color wheel and make surprising allies.

Color Pairings	Example
yellow and violet	firecracker flower and plumeria
orange and blue	Mexican flame vine and golden dewdrop
red and green	red ginger and prayer plant

To illustrate confidence and strength, choose three colors. They are triangles on the color wheel and, like all triangles, are stable and reliable.

Color Pairings	Example
red-yellow-blue	scarlet sage, lollipop flower, scrambling sky flower
violet-orange-green	yesterday, today, and tomorrow, gazania, papaya

To tie other colors to each other and make designs look and feel complete, choose other plants as well as pots and accessories in neutral shades from white to gray and brown.

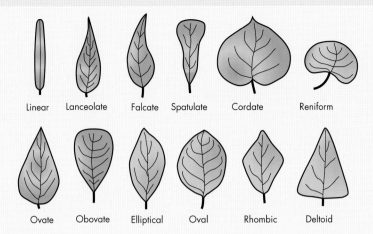

| Linear | Lanceolate | Falcate | Spatulate | Cordate | Reniform |

| Ovate | Obovate | Elliptical | Oval | Rhombic | Deltoid |

Leaf Shapes

Sophisticated and inspiring, leaves that are longer overall than wide bring a timeless sense of elegance to designs.

Shape	Description	Example
Linear	thin, longer than wide, pointed	Port St. Johns vine
Lanceolate	stiff tear drop or saber	rattlesnake plant
Falcate	soft tear drop	gold star, Persian shield

Friendly and welcoming, leaves with broad, entire surfaces can look coarse or quite soft but are never threatening.

Shape	Description	Example
Spatulate	wide tips form clubs	peacock plant
Cordate	hearts point away from stem	elephant ear, taro
Reniform	kidney or lima bean shapes	split leaf philodendron

Fat and sassy leaves with wide middles make a strong statement of permanence that cannot be ignored even in diverse plant groups.

Shape	Description	Example
Ovate	broad bases flare out to a point	patchouli, chocolate plant
Obovate	broad tips that taper in to a point	chameleon plant
Elliptical	football or sword shapes	dumb cane
Oval	broad, egg shaped	Madagascar jasmine

Angular and reliable leaves with standard geometric shapes stand erect like guards and can convey mystery or aloofness.

Shape	Description	Example
Rhombic	diamond	justicia
Deltoid	triangle	jade vine

Container Gardening
Tropicals in Pots, Baskets, and Boards

Growing tropical plants in containers has many advantages, primarily in creating and managing their conditions. Habitats and microclimates vary widely, even within frost-free zones and certainly

elsewhere, and gardeners often want to exceed those limitations. Lots of tropical plants can be grown as summer garden annuals in the temperate zones and replaced each year. But many others do not adapt to soil types and rainfall patterns that differ greatly from their comfort zone or do not mature quickly enough to be grown as annuals. For example, a garden where heavy (clay) soil stays cold well into spring can limit a tropical plant's potential regardless of its suitability otherwise. Or a plant that needs a dry spell to flower may disappoint the gardener where rainfall is evenly spread through the year, even if temperatures stay above freezing. By growing such plants in containers, their needs can be met more easily.

Soils and Potting Tips

The majority of tropical plants and all the ones profiled in this book can be grown in soil that is fertile, organic, and well-drained. Even top-shelf potting mixes may need amending with organic matter and coarse sand or other materials to ensure these conditions are met. It is important to know if and how much fertilizer and water-holding polymers are included in a potting mix to avoid overdoing these essentials. A proper growing medium will sustain nearly any plant, but there are exceptions among epiphytes, cacti, and orchids. For example, many cacti thrive in soils that are sandier than average, and some orchids can best be grown in a mix that's nearly all bark. In their natural environment, epiphytes grow in the ridges of tree bark, where water pours over them almost daily but never lingers for too long. In their case, the container becomes as important as the growing medium. A board or driftwood is needed for epiphytes such as staghorn fern and some bromeliads in order to mimic their native conditions of small pockets of soil. Other plants depend on the gardener to provide a suitable container, one that drains well and is big enough to hold its soil and roots. Except for some epiphytes, containers can be hanging or upright. Baskets are the most common hanging pots, but half pots and woven fabric or coir pockets meant to hang on a wall or fence deserve consideration. Their sizes vary, but like upright pots, a smaller volume often means a greater need for watering. Materials used to make pots vary, too, but a useful rule of thumb is this: clay pots and fiber baskets will dry out faster than plastic or ceramic pots of the same size.

When you are repotting plants, consider the container's material. Clay and fiber pots dry out faster than plastic and ceramic pots.

Potting plants successfully takes no real talent, just a gentle hand and a few basic guidelines. To protect roots from damage, water plants the day before their roots will be disturbed. It is a good idea to squeeze the outside of small plastic pots to loosen roots, but if a plant has been potted so long that its pot cannot let go, sacrifice the pot. Cut down one side of it to expose the rootball and ease it out of the tight pot. In most cases, a rootball that has white roots tightly wound around it should be roughed up by rubbing its sides or by making vertical cuts in its sides. Put the plant into its fresh soil and new container, leaving ½ to 1 inch of headspace between the top of the soil and the top of the pot. This practice allows deep watering without splashes and lost soil. Many potted

Crowded roots can choke off water uptake and slow or stop growth.

plants are watered best by filling their saucers with water regularly. This method is neat and efficient but should not include leaving water in the saucer for more than a couple of hours. In addition, these pots should be watered from the top once monthly to allow for air exchange in the soil. Unless a plant's leaves are very hairy, it will benefit from regular misting to deter insects and keep leaves clean indoors and outdoors in dry weather. Repot plants when their roots push out of the drain hole, when the pots tip over between watering, and when water rushes through the pots without time to percolate into the soil. Tropical plants can grow to be very large, but many will mature at a smaller size based on the available size of their root zone. When such a plant is as large as is desired, it can be removed from its pot and root pruned to control future growth, then returned to the same pot with fresh soil for another year.

Mixing Plant Needs

Although it is easiest to group plants by their needs for water, fertilizer, sunlight, and so forth, few gardeners do. Plants are chosen for their beautiful flowers, striking leaves, and sometimes their size and are expected to adapt, which some do. Others, however, have specific demands best addressed by housing them separately in containers. Regular applications of fertilizer will create lush perfection on one plant but another will be overwhelmed by too much nitrogen and never flower. By separating such plants into individual pots, they can grow side by side and still be treated singly. A plant that can steadily put on new leaves all year may be stifled if the light and humidity in its bed vary from season to season, but a potted plant is portable and can be moved to the best location each season.

Potted geraniums and lemon trees match up well in scale.

Pots for Style and Function

Not every reason to grow a plant in a container relates to gardening considerations. The gift of a spectacular glazed pot can be reason enough to start a container garden. Gardeners like to grow plants indoors and outdoors hanging from trees, poles, and porches. Even where space is not a consideration, a wall lined with planters makes a statement larger than its size, and indoors, one big plant sets a tropical tone for an entire apartment. Still other plants look best at one point in the year and deserve to be moved front and center for those weeks or months. In a small garden, the idea of rotating focal points by featuring containers in season can be very appealing as a way to add dimension and variety to a space. This can be as simple as picking the one in bloom from a group and moving it into better view, or it can be more complicated. Set up a pedestal where a plant can be viewed from several angles as a permanent part of the courtyard hardscape and move the candidate for "Star of the Month" into full view when its glamour is ready for close-up status. At other times, let it be part of the chorus line, less prominent but still integral to the scene.

The difference between a container garden and a group of pots is sometimes the pots themselves, but often it is how you put them together. Anything that can hold soil, drain water, and is big enough to hold a particular plant's roots can be its container. That includes wooden boxes, concrete tubs, recycled pop bottles, and cowboy boots. In practical terms, pots are usually either clay or plastic. Terra cotta clay pots have the advantages of high style and superb air drainage around root systems, which is helpful for plants that need to dry out between irrigations. Clay will crack in cold weather or with great age, but its material grinds up readily into a soil amendment useful for improving drainage. Plastic pots are durable, keep soil and root systems from drying out so quickly, and are molded into virtually endless designs to suit any taste. Plain black nursery cans come in every size, are lighter than clay, and happily take a backseat to their plants' beauty. Modern plastic garden pots come in a rainbow of colors and often mimic classic container designs, like urns and pedestal pots, but are lighter and much easier to handle than their clay equivalents.

Beyond the material used to make them, pots make themselves known by their colors and shapes. Decorative and functional, pottery and ceramic, huge and tiny, regular and self-watering, garden containers constitute a wide-ranging international market. With a few basic design principles in mind, any collection of potted plants can make a memorable scene. As in any garden design, a combination of wildly different flower colors and leaf textures can be too chaotic to really see and may be distracting. In the container garden, it is also possible for there to be too many different pot designs, or too many of one kind, for harmony to prevail. It is good to remember that less is often more. For example, a set of three nested, rectangular pots makes a chic addition to a set of steps or a low wall. They match in color and reinforce a chosen shape, and they step down in size to direct the eye with a sense of movement. Three sets, in different primary colors, can only look clownish in the same situations and will detract from whatever is planted in them. Likewise, three or five interesting pots will add to a group's drama, but ten distinct styles is just raucous, no matter how nice their plants might be. Should such a diverse collection be the issue, spread the wealth around by staging several displays, each highlighted by a few of the wild ones. To increase the number of plants in a group featuring a few bold pots, add plastic pots that will be almost invisible in the shadow of the stars.

Let pots set a garden's mood by staggering their sizes, shapes, and heights to create uplifting lines. Fill them with upright trees and vines on tall trellises to further the soaring scene, or soften the

Pots can influence a garden's mood by their size, height, shape, and composition.

lines with clumping plants and trailing types. At the other extreme, classic strawberry pots and other tubby types provide rounded, squat, quaint lines. They deliver a cozy, comfortable aura and beg not to be overshadowed by plants that are not suited for their wide, shallow shape. Pot shapes and ornamental designs offer lines that can complement or contrast with plants to enhance their impact or sadly detract from it. A clumping plant noted for its bizarre leaves is dramatic unless planted in a container more colorful than it is, or much bigger than need be.

A similar *faux pas* happens when a flowering plant finally opens its long-awaited blossom and its painted pot clashes gruesomely with the plant's natural colors. When it comes to the broader impact of colors in a garden, the same essential principles apply to pots as they do to plants. The cheerfulness of yellow flowers can be repeated in the color of their pots or reinforced across the garden in a container being used for another plant that repeats the signal color.

This ability to repeat and reinforce a color scheme makes pots a handy way to quickly establish unity in a garden design. By using a primary color only once in a design, that effect is reversed, as when a pair of bright red pots is used to mark the entrance to a garden room. When all other hardscape and accessories in the garden are shades of yellow and pink, those red pots work to set that space apart and call attention to it.

Grouping potted plants is done for their health and the convenience of the gardener. They can be grouped by their needs for sunlight or other requirements to make maintenance more routine and more likely to get

Move potted tropicals indoors before cold weather sets in.

done. It is admittedly harder to forget to water a group of plants than it is a single pot in the corner of the dining room. Indoors in climate-controlled environments and outdoors in dry climates, many tropical plants benefit from groupings designed to raise humidity. Actively growing plants increase

Water the soil, not the leaves unless they are dusty. By setting potted plants on gravel, you increase the humidity level when you water the saucer.

the humidity in the area immediately around them since they transpire (evaporate water) through their leaves. By placing several plants close together on a tray of gravel, each plant benefits from the increased humidity present in the space. Add a daily misting to the top of the gravel to evaporate more water and it is possible to keep leaves hydrated without overwatering the plants or wetting their leaves between waterings. Sometimes it is necessary to treat the water to accommodate plants sensitive to "hard" water, which is known to be rich in minerals. Many gardeners routinely allow water to rest in a watering can until it reaches room temperature before using it on tropical plants in pots.

Another smart way to facilitate container plant care and feature groups of potted plants in a garden's design is with effective staging. The classic garden staging device is an opened ladder with glass shelves laid across its A-frame to hold small pots. Such simple structures have a solid but not imposing design, and they emphasize utility over fancy details. But whether they are clean lines or scrolled in gingerbread curlicues, permanent or temporary, indoors or outside, it makes sense to use lifts, hangers, and shelves to put plants in their best light, literally and figuratively.

A great part of the allure of container gardening is in being able to grow plants out of season and out of place. Just as an expert can be said to be a journeyman far from home, a plant may be common in its native land yet considered exotic and prized in another zone. There is joy in bringing rainforest natives to areas far from the equator and in nurturing desert denizens in the land of monsoons. Such displaced plants can be more challenging to grow, and that is another aspect of their appeal to those who enjoy having a plant that few in the neighborhood share. The desire to collect plants from other climates can mean starting a traditional indoor garden with its range of possibilities and rewards. But in some cases, the tropical plant gardener may take other paths to see specimens through colder winters than they can survive unaided. An obvious strategy is to propagate the plants that cannot be dug up and moved indoors so their offspring can be returned to the garden the next year. Those who like to root plants often take cuttings in summer and fall from plants they will still overwinter inside as a way to increase what they can share and trade with others. Look for more information about multiplying plants in the chapter on propagation and in the individual plant profiles in this book.

Cool-Storage Strategies

A glass conservatory with automated fans and heat to keep tropical plants happy may be the secret dream of every gardener, but it is one few will attain. The space, maintenance, and expense of running a greenhouse can be daunting, but they operate on universal principles. Everyone who lives outside the tropics can use some of the same strategies to overwinter plants outside their comfort zone. Tropical plants fall into three broad categories for these purposes: very tender, tender, and sometimes hardy, often called half hardy. The first group is hardy only in the true tropical zones and will succumb to temperatures near freezing and certainly below it. Very tender plants must spend the winter in a warm environment, well above freezing, or they will be damaged. Most of the traditional tropical foliage plants fall into this category and they adapt well to a traditional indoor garden setting. The second, tender plants, are hardy in the subtropics as well as the tropics, either as evergreens or returning perennials even farther north. Some, such as batface cuphea, can be reseeders too. Some tender plants have the potential to tolerate subfreezing temperatures, but only for a short period. They can be added to the indoor garden and some can be protected outdoors quite effectively in warm winter areas. Think of it as cold storage for containerized trees, shrubs, and vines that are woody; those that can go dormant; and big pots that can be cumbersome to bring indoors. By protecting the roots primarily and letting the top growth suffer a little, it is possible to carry plants over from year to year as far as two zones farther north. Half hardy plants can be brought indoors or given a similar cold storage setup to protect them even farther north. As outlined in each plant's profile, the plants that are good candidates for storage outdoors are able to regrow from their roots, as long as those roots are not frozen or completely dehydrated.

To establish cold weather protection for potted plants, determine where they can be housed. (The same kind of shelter should be given to pots that might crack if frozen, even if they are empty in winter.) The site must be out of the weather but not necessarily heated or lit well. A garage, barn, potting shed, or A-frame attached to a house will work. Gather bales of hay or pine straw, or select another insulating material. Cut the plants back that are to be stored, those that are marginally hardy

Many plants can be overwintered inside your home such as in this sunroom just by bringing their pots indoors.

where you live and/or will go dormant, so they will have about half as much top growth to support as they did in the growing season. Propagate all plants that can be rooted as insurance against winter damage. Put the plants to be stored inside the shelter. Line them up pot-to-pot and fill in and around with mulch, then position the hay bales around the entire group. Water much less often and, if it is possible, open doors and windows on sunny days for air exchange and added light. Enclosing porches and carports with plastic sheeting improves the environment where it is practical, and if a window can open into the heated house, so much the better. Give in to curiosity and put a thermometer near stored plants, preferably one that records nighttime lows and daytime highs. If supplemental heat is desired, take care to use only appliances registered for outdoor use and follow all directions for safe use, including ventilation requirements.

Bringing Tropical Plants Indoors
Breathe Easier

Once you give in to the way tropical plants can soften any architecture and bring life to indoor space, their absence leaves you feeling chilly. A lush palm tree next to the couch in a physician's office warms the room and has a positive effect on the waiting patients. Its presence instantly communicates a feeling of tranquil good health that may make the time seem to go faster. Western European explorers and their patrons were so hypnotized by tropical plants, flowers, and fruit that heated glass conservatories were invented to accommodate their newfound passion. Since then, the popularity of

indoor plants has waxed and waned. When their star shines, it is at times widespread enough to inspire furniture, such as fern stands for Victorian parlors and whole armies of macramé plant holders. Fortunes are built on the most beautiful, useful designs and materials for plant containers, and gardeners prize their favorites. Flowers, leaves, and even trees and vines have the same warming effect on homes simple and grand. Lovely as these benefits are, growing tropicals has tremendous, measurable impact on the quality of the air indoors.

Indoor Air Quality

At least since the Industrial Revolution, people have known that indoor air could make them sick. As manufacturing grew, the kinds and concentrations of chemicals used in enclosed spaces grew too. From nineteenth-century hatters and printers to asbestos workers, fumes and particulates have been inhaled by workers to their detriment. The issue of indoor pollution began to gain steam in the 1970s, when architectural trends toward tightly sealed buildings met rising awareness of the hazardous potential of chemicals used in manufacturing a host of building products. From wallboard to rugs and beyond, volatile organic chemicals, particularly formaldehyde, were found in the end products and so, in living spaces.

Modern Solutions

As early as 1982, NASA identified more than a hundred volatile organic chemicals (VOCs) being emitted by materials in Skylab III. Clearly, the closed environments of space would benefit from the cleanest air possible, and phytoremediation—using plants to filter pollutants—has proven its worth. NASA scientist Dr. Bill Wolverton began to measure the effects of common indoor pollutants on houseplants in the 1980s at the Stennis Space Center in Bay St. Louis, Mississippi. He enclosed individual plants in sealed plastic chambers and then injected each chamber with common pollutants and measured the results. Dr. Wolverton set the standard for measuring the ability of plants to clean polluted air (the phytoremediation) and gave us all an excuse to expand the indoor garden—as if we needed one. This groundbreaking work continued after he left NASA and has included important books on the subject such as *How to Grow Fresh Air* and *Plants: Why You Can't Live Without Them*. See the Resources section at the end of this book for more information and references.

Groups like the National Foliage Foundation and researchers from around the world have added to the case in research reported in 2011. Indoor plants remove carbon, VOCs (like benzene and formaldehyde), and carbon dioxide at rates that rival mechanical filtering systems, and more plants are efficient at this process than previously known. Project Carbon at the University of Georgia proved that yes, plants remove carbon from the air if they are healthy, and that larger, woody plants absorb and retain more carbon for longer periods than green-stemmed plants. Research by UGS horticulturists with scientists from Korea's Rural Development Authority reported on the benefits of eighty-six plants to remove formaldehyde effectively from indoor environments. Their ratings found that Japanese royal fern (*Osmunda*) and other ferns proved among the most effective tested. NASA's list of the most effective plants for this purpose includes some featured in this book: bamboo palm, gerber daisy, snake plant, and several *Draceana* species. Other tropicals on the "A" list are Chinese evergreen (*Aglaonema*) and peace lily (*Spathiphyllum*).

More Concerns

As the research began to accumulate, better products came to the market, yet disturbing levels of harmful pollutants can remain where we live and work. In 2002, a World Health Organization report estimated that these "indoor volatiles" are responsible for more than 1.6 million deaths each year, or 2.7 percent of the global disease burden. Admittedly, one gardener's collection of houseplants, however extensive, can only clean the air in that one space. But by encouraging others to follow suit and supporting efforts that bring indoor gardens to hospitals and other facilities, clean air can become the norm.

The spectrum of visible light is ROYGBV: red, orange, yellow, green, blue, violet.

More indoor garden plants succumb to overwatering than to anything else, but inadequate light runs a close second on the list of fatal flaws. The problem can be the result of too much or too little light either in terms of its intensity or its duration, or both. Signs of low-light stress on plants include pale leaves, stretched stems, small or nonexistent flowers, and very slow or no new growth. Too much light produces its own set of issues in plants, such as burned leaves, rapid soil drying, and growth that is oddly too compact for the species, sometimes including flowers at the expense of leaf growth. Variegated leaves react too. Low light can cause them to revert entirely to green, while excess light can extract the green, leaving only yellow or white. A quick change in the available light can cause leaves to drop from an otherwise healthy plant and is exacerbated by a radical temperature shift. That is why gardeners outside the frost-free zones are well advised to transition pots *gradually* between indoor and outdoor locations. Move pots from full sun into a partly shady site outdoors for two weeks before bringing them inside, even to a well-lit room. To avoid abrupt temperature changes, finish the transition at least one week before it will be necessary to heat the inside of your home. To best accomplish the move, determine the date of first frost in your area and begin moving the most tender plants to lower light outside about six weeks before then. The key to proper light for gardening indoors is knowing how much light plants need, how much is available, and what strategies will work best to bring more or better light if needed. Often an approach that combines natural light with artificial, or supplemental lighting proves best, and modern options make it as efficient as it is effective to do.

Light Works

Schoolchildren learn about the spectrum of visible light with the acronym ROYGBV, which names the colors in order of frequency: red, orange, yellow, green, blue, and violet. Light is what you see, part of the electromagnetic spectrum that also includes non-visible elements like infrared. Anything in your sight line is absorbing some colors of the light spectrum and reflecting others. It takes the full spectrum of visible light to grow plants, but we see only what they reject. When leaves look green, it is because they are reflecting the green portion of the spectrum to our eyes. Few plants look blue or purely red because those colors are absorbed to play a role in leaf and flower development, respectively. When you understand how light and plants relate to each other, you can make better choices and grow the indoor garden to its beautiful potential, full of green air cleaners and spectacular flowers.

The intensity of a light source and its duration combine to sustain thrifty growth or stifle it. Intensity measures two factors: how bright the light is and how far away it is from the plants. The strength of the light source itself is reported in lumens, as you would see on a package of light bulbs. Footcandles measure the light that falls on an object from a specific distance. For example, 100 foot-candles of light will enable the average person to read a newspaper without squinting but will sustain only the most shade-loving plants. By the same turn, you might need sunglasses to read in 1,500 foot-candles, the amount needed for orchids to bloom. How long the light lasts plays an enormous role in plant survival, but it's even more important in promoting their growth. While the actual length of a day does not change, the number of hours of sunlight (or supplemental light) varies. By mimicking the variance in a plant's natural light cycles, you can alter its bloom cycle. Long days and short nights (6 to 8 dark hours) trigger many mature bromeliads and hibiscus to flower, as happens naturally in

The key to proper light conditions indoors is to know how much light plants need and whether natural light is sufficient.

summer in the Northern Hemisphere. Many orchids, along with gardenias and some *Euphorbia* species, bloom naturally when darkness extends for 12 hours in a single cycle.

More Light

Adding supplemental lighting to a room used to be an ugly affair, limited to garish bulbs, wire shelves, and clunky fixtures with all the style of a fluorescent shop light. That two- or four-bulb, aluminum-hooded thing from the garage is still around, of course, and remains a workhorse. You know it is on because it hums, vibrates, and soon flickers annoyingly as its ballast is exhausted. However, armed with equal numbers of pink daylight and cool white bulbs, it provides a nearly perfect rendition of the full spectrum of light needed to grow plants.

Improved technologies in lighting have come a long way, baby, but fluorescent bulbs remain the standard for most indoor gardens. The best in plant light bulbs deliver a much more complete light spectrum and are much more energy-efficient than previous bulbs. They are compact, have a lower profile, last longer, use as much as 20 percent less electricity, and produce much less heat. Modern fixtures are better, too, and are widely available in sizes to suit any space. They have solid state ballasts and do not hum like the older models. Perhaps best of all, newer lights have less glare without those huge, hooded fixtures and provide better light for people as well as plants. Spiffy-looking LED grow lights have great potential to transform dark corners everywhere into plant heavens. The fixtures are small with adjustable heights and long-lasting panels of LED bulbs that are blue, to assist photosynthesis. High-intensity lights produce copious amounts of full spectrum light, but are best suited for large or very dark spaces because they are expensive to operate. Metal halide lamps produce the most light and fullest spectrum of the high-intensity types, while high-pressure sodium lamps have very bright light for blooming plants.

Dos and Don'ts

You can make the most of any light source, natural or artificial, by designing the space and maintaining it to capture and reflect all available light. Add a well-placed mirror, paint a wall white or another very light color, and use sheer curtains that shield you from view without blocking the light. When remodeling, replace small windows with larger ones, and consider taking down some window screens when appropriate to let more light into the room. In any case, wash your windows regularly to maximize light. Some light types are simply unsuited for growing plants. Incandescent light bulbs, even the energy-efficient types, lack the full spectrum of light and produce more heat than most plants can stand. The heat issue and their narrow design keep halogen lights off the list, even though they do have a wider spectrum.

Flowering Plants Indoors

It can take many hours of high light for tropical plants to bloom indoors, but the best choices pay off handsomely. There are no hard and fast rules, but good candidates will have relatively thick leaves

Jade plant (Crassula ovata)

and moderate needs for water. It bears repeating that more plants die indoors from overwatering than low light, pests, or underwatering. Nearly all plants require less water indoors than they do outside, even when supplemental lighting is being used. Even water-lovers benefit from more time between irrigations. In general, let the soil surface look a little dry, or pick up a small pot to see if it feels light to know when to water indoor plants. Add water to the saucer under the plant to keep the leaves and flowers dry but remember that good air exchange in the soil requires watering from the top at least occasionally. Keep slick leaves clean by simply wiping each one with clean water on a damp rag, but avoid getting water on hairy leaves at all.

Humidity

With light and water management well in hand, there is one more factor to consider in growing tropical plants indoors. Humidity is the measure of the water present in an air sample, carefully kept at human comfort levels around 20 percent by many heating units. While many plants can easily tolerate lower relative humidity indoors, most tropicals want more, about 50 percent. The easiest way to increase the humidity around plants is to include plants with big leaves in every grouping. The humidity is higher around big-leaved plants because leaves evaporate water as they transpire, and big leaves transpire more than small ones. A tray of wet pebbles under the pots can increase humidity without wetting the plants at all, as will humidifiers incorporated into heating systems or used as standalone appliances.

Indoor Microclimates

Just as the native habitats of tropical plants are separated into microclimates, so is your home, office, or other indoor space. Depending on its orientation, architecture, and purpose, each room defines a zone of available light, average temperature, and humidity that can be quite unlike others in the same building. Obviously, supplemental light changes the microclimate, as does added humidity, giving you the opportunity to create the microclimate that best suits your collection. Sometimes these microclimates maintain the same growing conditions all year, but not always. For example, a table in front of a picture window that faces east might be ideal for flowering plants in winter, until the tree outside gets its leaves in spring. Then that table offers shelter for foliage plants or a flat of cuttings.

An overlooked asset of growing plants in containers is their portability, which allows you to move them about to find the best location for a particular plant or group. While each home is different, most share a set of features that can affect indoor gardening, and while plants are more adaptable than it might seem, some can be counted on in particular microclimates.

Kitchens and bathrooms. Clearly more humid than most indoor spaces, these two rooms have pros and cons for plant growing. They can be easier to ventilate than other rooms, a decided plus, but bathrooms may be poorly lit unless modifications are made to supply the full spectrum of light. Where light is ample, bathrooms are great places for ferns, coleus, and crotons. Well-lit kitchens, especially those with gas stoves, can cause plants to accumulate oils, grease, and dust on their leaves. It is wise to grow thicker-leaved plants in the kitchen so they can be cleaned easily. Make space there for blunt-leaf peperomia, dragon tree, and African milk bush.

Indoor spas and pools. Everyone should have lady palm, ferns, nun's orchid, and flamingo flower in these hot, humid indoor microclimates. Build a collection that is larger than you can use at any one time and rotate plants between this saturated atmosphere and a warm room with less humidity.

Cool spots. Areas like foyers, enclosed porches, and rooms on the north side of the house that have few windows are usually cooler and sometimes drier than the rest of the house. If you have a second climate control unit, those areas can create this microclimate and save energy, too. Cool spots work well for cactus family members, bush lily, and bird's nest fern, so long as they are not cold and drafty.

Windows and warmth. The orientation of your home and the windows in it almost automatically create microclimates that have two parts: close to the direct sun in the window and farther away, where the light is reflected. South-facing windows are too warm in summer for most plants but are useful in the winter, when their relative warmth is appreciated. Windows that face north offer a summer retreat for plants that would be sunburned by the south exposure. In winter they can sustain woody plants in pots that need resting, such as shower orchid as it prepares for late-winter bloom.

Vents, doorways, and fireplaces. If you gave a tropical plant a list of hazards to be

Humidity and light levels in bathrooms have pros and cons to growing tropicals such as this orchid.

avoided indoors, these three would top it. The area beneath or next to a vent is usually drier than elsewhere in the room and blasts cool or warm air regularly. Doors to the outside that are used frequently can expose plants to the elements unnecessarily. Fireplaces and space heaters are hot, dry properties not well suited to most plants.

Labor of Love

Beyond clean air and good looks, indoor gardening is personal horticulture therapy, the home version of that venerable scientific discipline. The pleasure of caring for plants can and should be just that—a relaxing and enjoyable experience. But gardening has specific benefits that are well recognized and used internationally in therapies to foster self-esteem, control anger, improve critical thinking, and address other personal issues. Many of us see the garden as our safe nest, where we

can let our hair down and rest from the worries of the world. When that attitude is applied to therapeutic settings, people who never thought they would be gardeners take to it like ladybugs to aphids. It can be enlightening to realize that, unlike most of life, a garden is in your control. You figure out what a plant wants, provide it, and the plant grows. The realization that you did that, that you kept a plant alive, is pleasing and can be immensely gratifying for someone facing tough times in other areas of life. Let's face it, when a flower blooms in your very own living room, it feels like a pat on the back. Perhaps best, plants never talk back and, unlike relatives and coworkers, they do come with instructions.

In certain zones, large ferns (Nephrolepsis) *such as these can be planted in the landscape.*

The Best Indoor Tropical Plants for Cleaner Air

The best tropical plants to grow indoors are those that thrive in conditions usually best suited for humans: less-than-full sunlight, moderate temperatures, and moderate humidity levels.

Besides these general criteria you can take advantage of NASA's research, which focused on the ability of plants to clean the air indoors. Choose among these high-performing plants to clean the air in an average home. For every 1,800 to 2,200 square feet, grow 15 to 20 potted tropicals that are 6 inches or larger in diameter. These plants come from shady places in the tropics, are best adapted to indoor gardening, and are widely available everywhere.

Plant Name	Botanical Name
Dracaenas	
Red edged	*Dracaena marginata*
Warneck	*D. deremensis* 'Warneckii'
Janet Craig	*D. deremensis* 'Janet Craig'
Cornstalk	*D. fragrans* 'Massangeana'
Philodendrons	
Heartleaf	*Philodendron scandens* 'Oxycardium'
Elephant ear	*P. domesticum*
Selloum	*P. selloum*
Airplane plant (spider plant)	*Chlorophytum comosum*
Bamboo palm	*Chamaedorea sefritzii*
Chinese evergreen	*Aglaonema modestum*
English ivy	*Hedera helix*
Peace lily	*Spathiphyllum* 'Mauna Loa'
Snake plant (mother-in-law's tongue)	*Sansevieria trifasciata*
Weeping fig tree	*Ficus benjamina*

In addition, NASA focused on 3 common household pollutants often found in building materials and household furnishings. Here are some of the results showing plants and the chemical they clean. It's just a little more motivation to grow tropical plants indoors.

Plant	Benzene	Formaldehyde	Trichloroethylene
Airplane plant		x	
Bamboo palm	x		x
Dracaena		x	
Gerber daisy cut flowers			x
Peace lily	x		x
Pothos ivy		x	
Snake plant	x		

Source: Cleanairgardening.com citing NASA Wolverton studies

Herbaceous Tropicals

African Mask

Alocasia × amazonica

African mask is beloved for dark green, deep V-shaped leaves with ribs and edges marked in bright creamy white. Jaunty and full of attitude, it laughs off legendary arguments about its parentage and lack of official standing by many in the scientific community. This colorful plant is neither a species nor found in any rainforest, including those in Southeast Asia where its relatives are native. Experts have traced its family tree back to Amazon Nursery, which was owned by Miami postman Salvatore Mauro in the 1950s. Whether it is a hybrid or a stable offspring of another Alocasia matters less than its ability to deliver distinctive tropical flair. Outdoors, African mask haunts the shade while indoors, any bright room will do for this long-lived plant. Aroids, as plants in this family are called, display rowdy, outrageous leaves, especially this rogue member.

Plant Family
Araceae

Other Common Name
Alligator plant

Bloom Period and Seasonal Color
Yellow flowers appear sporadically amid boldly patterned leaves on older plants

Mature Height × Spread
Up to 3 ft. × 2 ft.

When, Where, and How to Plant
Outdoors, select a site that offers light but is shaded from direct sun all day. A tabletop near a bright window amid a group of other plants will provide ample light and humidity inside a home or office. Prepare the garden soil or a container growing mix that is richly organic yet well drained. Use ground bark, compost, or compost/manure to enrich soils. Grow African masks away from wind and salt spray to avoid shredded and pocked leaves. Plant clumps 1 to 2 feet apart in beds. Repot when roots are visible in the drain hole, when water rushes through without percolating into the soil, or if the pot tips easily.

Growing Tips
African mask plants that are allowed to dry out will go dormant. Reservoir pots can be filled on *your* schedule, and drip irrigation or garden bed soakers can be automated to ensure good hydration. Fertilize moderately using a granular formula three or four times annually, or use a soluble fertilizer mixed at half strength every other month. Keep organic mulch around plants in beds and pots. Its leaf tips may brown if the humidity is too low or if your tap water is unsuitable.

Care and Propagation
Keep an eye out for spider mites, mealybugs, and scale insects. To propagate, divide rhizomes while they're dormant or separate offsets from the mother plant. See "Propagation" (page 36) for details.

Companion Planting and Design
African masks can hold center stage alone or surrounded by a bed of nerve plants. Pair it with striped blushing bromeliads, golden brush gingers, or ruffle ferns for a bold tropical vignette.

Try These
Dwarf varieties and selections have confusing names too. For example, *Alocasia* 'Poly' is also known as 'Polly'.

African Milk Bush

Euphorbia umbellata

When, Where, and How to Plant

Choose or create a sunny, dry site for these plants, such as areas appropriate for cacti or succulent gardens. Plant it near walls or other plants that can act as a warming windbreak, especially in the temperate zones. African milk bushes thrive in very well-drained, fertile soil and an environment of low humidity. Outdoors, prepare a bed of existing garden soil amended with sand and organic matter to meet these needs. This plant grows well in containers with potting mix that has ground bark or sand mixed into it to ensure excellent drainage. Avoid mixes with water-holding polymers that can keep the root zone too wet. Space plants for excellent air circulation in beds or grow one in a standard 10- to 12-inch clay pot. Do *not* mulch these plants.

Growing Tips

Allow African milk bush plants to dry out thoroughly between waterings when not in bloom. When flowers appear, water the plants regularly but not excessively. Keep plants groomed by removing old leaves and flowers once they have dried. Fertilize four to six times annually. Use a granular or soluble complete garden fertilizer. Repot annually to provide ample room for growing roots. Move pots into a sunny, warm room for the winter.

Care and Propagation

It has few pests. Propagate African milk bush by taking cuttings.

Companion Planting and Design

Grow African milk bush plants with other *Euphorbia* plants such as crown of thorns and succulents such as desert rose and Cape aloes. Let one take center stage surrounded by devil's backbone, gazanias, and bushy hibiscus.

Try These

'Rubra' is widely planted for its red stems and leaves, as well as its flowers.

African milk bush adapts well to a variety of environments in the tropics and subtropics and in sunny indoor spaces worldwide. Indeed, its other common name, chameleon plant, might refer to its habitat flexibility. Native to East Africa, it is named for the white latex sap exuded by its stems and leaves when they are broken. Like many other Euphorbia species, the sap can sometimes be a skin irritant. The leaves jut out stiffly from chunky stems like stacks of bright green helicopter rotors. Their veins are a network of darker stripes on each leaf. Red stems spring from the top of each stem, slick, stocky, and exposed like latticework. Each forked stem holds a bright red flower that looks like a fancy summer straw hat hung in a milliner's display window. There are dozens of flowers atop each stem, the perfect crown for such a beautiful succulent shrub.

Plant Family
Euphorbiaceae

Other Common Name
Chameleon plant

Bloom Period and Seasonal Color
Red flowers in the spring and summer

Mature Height × Spread
Up to 12 ft. × 2 ft.; up to 6 ft. x1 ft. in containers

Angel Wing Begonia
Begonia hybrids

As easy to grow as it is to love, angel wing begonia lives up to its name. The leaves are wing-shaped, sometimes elongated almost ridiculously to climax in deep points. The rounded top leaf edges are elegantly draped, sometimes cupped or serrated along the edges, and often splattered with accent colors. They look slightly bizarre but effortlessly charming. Angel wing begonia leaves are many shades of green but also bronze, and shades of red spread warmly on the undersides. Clusters of red, white, and pink flowers are borne in the leaf axils on waxy stems to match. Each inflorescence has a pendant-shaped bract that opens like a locket to reveal tiny white flowers. Soon the seedpod forms and hangs below the bract in a stunning display. Native to Brazil, this begonia is at home in the shady tropics and sub-tropics everywhere and can summer outdoors each year in temperate climes.

Plant Family
Begoniaceae

Other Common Name
Cane begonia

Bloom Period and Seasonal Color
Red, pink, or white flowers off and on through the year

Mature Height × Spread
1 ft.–4 ft. × 1 ft.–4 ft.

When, Where, and How to Plant
Grow angel wings in shade or part shade outdoors or in a bright, humid indoor space. The angels grow best in richly organic, well-drained soil in garden beds, containers, or window boxes. Amend garden soils and potting mixes with a combination of organic matter such as compost and ground bark. Test a pot of the mixed soil by watering it well to be certain it will drain promptly. If it does not, add more ground bark. Plant angel wing begonias in pots 6 to 8 inches wide and deep, or space them 8 inches apart in garden beds. Pot up garden plants and move the containers indoors for the winter. Angel wing begonias may drop their leaves indoors in the winter, indicating their need for more light and warmth.

Growing Tips
The jointed, bamboo-looking canes of angel wing begonias will collapse if subjected to drought. Water regularly and fertilize monthly with a complete, balanced formula product. If the flowers are few, even in part shade, use a flower formula fertilizer in the winter and spring. Deadhead flower clusters when they fade to stimulate new growth and more flowers. Take cuttings in the fall from garden beds and grow the young plants indoors during the winter.

Care and Propagation
Watch for mealybugs and aphids on potted plants, and snails in the landscape. It can be propagated very easily by stem and tip cuttings. See "Propagation" (page 36) for details.

Companion Planting and Design
Fill a shady bed with angel wing begonias instead of impatiens or New Guinea impatiens. Combine it in pots with caladiums, rattlesnake plants, and bird's nest ferns.

Try These
Try 'Airy Fairy' for its dainty cut leaves and 'Avalanche' for its white arty splotches.

Bamboo Orchid

Arundina graminifolia

When, Where, and How to Plant

Provide a soil that is rich in organic matter and very well drained. Amend garden soils and potting mixes with organic matter in amounts equal to one-half to one-third of their volume. For best drainage with enhanced fertility, combine coarse orchid bark and sand. The plants will tolerate some shade, but for best flowering, choose a sunny or partly sunny site. Grow bamboo orchids in beds or containers outside, and in a warm, bright room indoors in cold weather. Plant a clump of 3 to 5 canes in a 12-inch container or space clumps 1 foot apart in warm garden soil. The clumps multiply readily and spread to fill pots and garden space.

Growing Tips

Very regular water is essential, but puddling or flooding may rot the canes. Fertilize with a granular complete formula product twice each year, in the spring and again in summer. Nurture the clump by working in mulch as it rots or adding organic matter to the soil regularly. Cut the flower stems down to soil level when blooms are finished. Repot when water rushes through the pot because of crowded canes. Bamboo orchids are forgiving of light freezes but can become weedy in the tropics.

Care and Propagation

Luckily, there are few pests to bother bamboo orchids. Propagate by dividing the clumps when they're not in bloom and by removing the keikis—plantlets that form at the nodes or along the flower stem.

Companion Planting and Design

Since bamboo orchids can tolerate some shade, grow them under a vine like Mexican flame or golden trumpet. Add height by planting them with yellow cestrums or night-blooming jasmines.

Try These

Bamboo orchids are generally labeled as such but look for slight variations in flower color.

Imagine what might happen if a bamboo plant seduced a classic orchid. Their offspring would be reedy and grassy with incongruous flowers hanging in the spaces between leaves. Bamboo orchid does exist, and it is a true orchid seen growing wild by all who visit the volcanic slopes on the Big Island of Hawaii, where it has become an invasive exotic plant. Fortunately for gardeners in less-ideal climes, it is a fast-growing, clumping terrestrial orchid that looks fragile but is not. Narrow, medium green leaves about 5 inches long stand out from thin stems with precious flowers 3 inches wide or slightly larger. Arranged in sprays of five buds each, each flower has three wide, pointed white or pink sepals above the flower tube and lip in purple shades with a yellow throat. These Myanmar natives thrive in sunny tropical gardens but readily bring their casual elegance to containers and garden beds everywhere.

Plant Family
Orchidaceae

Other Common Name
Volcano orchid

Bloom Period and Seasonal Color
Fragrant flowers are solid white or white and purple with painted lips that appear from summer to fall

Mature Height × Spread
Up to 2 ft. × 1 ft.

Bat Face Cuphea
Cuphea llavea

If some plants knew the names that humans give them, they might not like it one bit. For example, this dynamic sun-lover blooms incessantly in jewel tones of purple and red, attracts wildlife and, in well-drained soil, can be perennial into the 25 degree range and beyond (with protection). Instead of a name that celebrates its rugged stance and heat tolerance, it is named for a feature best seen with a squint and perhaps a wink. Bat face cuphea is a hummingbird magnet, with scores of 1-inch-long purple flower tubes with two lips facing up to welcome them. In someone's imagination, the tube tips became a bat's face and the lips became ears; the whimsical name stuck. Bat face cuphea plants are thick and rather shrubby at times, with dozens of stems full of green or gray-green pointed leaves.

Plant Family
Lythraceae

Other Common Name
Batface

Bloom Period and Seasonal Color
Red and purple flowers in the summer and fall primarily; everblooming in frost-free areas

Mature Height × Spread
Up to 2 ft. × 2 ft.

When, Where, and How to Plant
For summer flowers on bat face cupheas, start seeds indoors in late winter or purchase small plants in the spring. Choose a site in sun or mostly sun with a reliable water source nearby. Prepare a soil in garden beds or containers that is richly organic and fertile with very good drainage. Space bat face cupheas 8 inches apart in garden beds and mixed planters, or plant three small plants in a standard 12-inch pot. Count on it: groups of three, planted as triangle, deliver much more visual impact than three in a row. These plants flourish in garden soil in warm weather and are not difficult to maintain over the winter as potted plants in the temperate zone. Mulch beds and pots with 1 to 2 inches of organic material.

Growing Tips
Water and fertilize bat face cupheas regularly and let the soil dry out only slightly between watering. Wilted plants may be stunted and fail to bloom. Maintain organic mulch around the plants and work it into the soil as it rots. Use a general-purpose garden fertilizer as often as directed from the spring through fall. Bat face cupheas may become perennial in the subtropics and pots can be held in cool, though not freezing, temperatures over winter. Mulch plants well in the fall.

Care and Propagation
Watch for slugs and snails on young plants. Propagate by cuttings and seed.

Companion Planting and Design
Bat face cupheas combine well with other red and purple flowers. Grow some with chenille plants, summer snapdragons, and fan flowers, or Chinese hat plants and golden dewdrop.

Try These
Look for 'Georgia Scarlet' and 'Tiny Mice'. Cigar plant (*C. ignea*) is a larger cousin.

Bird of Paradise

Strelitzia reginae

When, Where, and How to Plant

Select or create a sunny environment with a reliable water source nearby to grow bird of paradise. These plants can grow well outdoors year-round in the subtropics; elsewhere they can be grown in a bed and then be potted up for the winter, or spend all year in pots. Make room for their wide, shallow root systems in the landscape; site the birds away from paved paths to prevent damage to either. They thrive in a soil that can be watered regularly without becoming saturated. Prepare a soil for beds and pots that is richly organic, fertile, and well drained. Space plants 2 feet apart in garden beds and mixed planters, or provide a pot 12 inches deep and wide for one plant. Dig a wide, rather shallow hole and nestle the plant into its soil for stability. Lay a mulch blanket in beds and pots 2 inches deep with organic material such as shredded or ground bark.

Growing Tips

Water and fertilize bird of paradise regularly from the spring through fall. Let the soil dry out only slightly and always water before fertilizing. Maintain 1 to 2 inches of organic mulch, such as shredded bark, around the base of the plants and work it into the soil as it rots. Select a general-purpose garden fertilizer with a complete, balanced, slow-release formula.

Care and Propagation

There are few pests to bother bird of paradise. Propagate it by suckers and root divisions.

Companion Planting and Design

Grow with Chinese violets to its rear and Persian shield around its base for a stunning color palette. Use it to step down from red Abyssinian bananas along with yellow cestrums and bamboo orchids.

Try These

S. nicolai, the white bird of paradise, is a larger plant with striking white flowers.

Bird of paradise joins lobster claw and hibiscus in the pantheon of the most easily recognized tropical flower forms. Its potent silhouette is geometric heaven, assorted triangles sharply drawn by petals and bracts to create the characteristic "bird" with its plumed feathery headdress. Each flower has a banana-shaped bract that is green with purple overtones, huge yellow or orange sepals, and royal blue petals called a "tongue." They are held above large, dull green leaves on strong stems. The leaves form a quirky, cheerful clump by displaying all the leaves in one geometric plane. This South African native enjoys popularity well beyond the tropics and sends its massive flocks of flowers soaring to florists worldwide. In the temperate zone, bird of paradise appreciates a warm wall to its rear, all the better to cast its long and familiar shadow.

Plant Family
Strelitziaceae

Other Common Name
None

Bloom Period and Seasonal Color
Orange, rose, and purple flowers year-round

Mature Height × Spread
Up to 3–5 ft. tall x 3–5 ft. wide in warm climates

Bird's Nest Fern

Asplenium nidus

Bird's nest fern may be the perfect emissary for the fern family as a whole. Light green and stiffly erect from the ground up, this plant lifts itself up in a wide vase shape to be admired. Its species name, nidus, means "nest" and connotes its friendly shape perfectly. Each frond emerges as a tall crook from a flat, dark center and unfurls optimistically, shiny and gleaming, ready to join its brethren. Unlike some other ferns, bird's nest adapts very well to indoor environments. The fronds are stiff and thick compared to most ferns and hold themselves dramatically upright in low light. The plants are endlessly interesting to their gardeners, some fronds nearly straight and sword-shaped, others as curvy as rickrack on a square dancer's skirt. Bird's nest fern is native to tropical Asia, where it grows as an epiphyte across a wide geographic range from Japan to Australia.

Plant Family
Polypodiaceae

Other Common Name
None

Bloom Period and Seasonal Color
Shiny leaves, primarily light green, but also in darker shades, most with dark midribs all year

Mature Height × Spread
Up to 5 ft. × 3 ft. in the tropics; averages 1 to 2 ft. × 1 to 2 ft.

When, Where, and How to Plant
Avoid direct sunlight unless the plant is acclimated when it's young. Select a site in shade or mostly shade with bright, diffused light and ready access to water. Prepare a richly organic soil that drains very well by amending garden soils and potting mixes. To the existing soil or mix, add compost and finely ground bark equivalent to one-half of its volume. Mix these and water well once to test its drainage, then add more organic matter if needed. Plant one bird's nest fern in a pot 6 to 8 inches wide or mass three in a planting space 15 inches wide.

Growing Tips
Recreate a bird's nest fern's natural environment by watering very often but maintaining a well-drained soil. Fertilize young plants monthly and older ferns four times annually. Use a granular or soluble product with a complete, balanced formula. Keep plants mulched. Repot when water runs through the pot too rapidly for any of it to be absorbed. Move it up to a pot 1 inch bigger around, but no larger, or the plant will spend all its energy growing new roots.

Care and Propagation
Watch for mealybugs and scale insects. Brown leaf tips is a sign of overwatering, while bronzing implies too much sun, too soon. Propagate this plant by divisions.

Companion Planting and Design
Bird's nest ferns have strong upright lines that add interest to shady beds and pots with impatiens and red gingers. Add pots of bird's nest ferns to humid bathrooms to decorate your home.

Try These
Asplenium antiquum, Japanese bird's nest fern, can grow outdoors in parts of zone 8. *A. bulbiferum* has lacy, 3-foot-long fronds that curve downward.

Blue Butterfly Flower

Rotheca myricoides

When, Where, and How to Plant

Blue butterfly flower thrives in organic, fertile, well-drained soil. In garden beds and containers, the soil must tolerate being watered very regularly and remain moist in between, yet continue to drain well. Amend garden soils and potting mixes with organic matter as needed to achieve this condition. Plant blue butterfly flowers 8 inches apart in beds and large planters, or provide a standard 8-inch pot for one plant. Unlike some of its relatives, this plant makes a good mate for mixed flower groups and is not a bully. Flowers are more abundant in full sun, but butterfly flower grows and blooms in dappled sun. Mulch plantings with 2 inches of organic material such as ground bark; use 1 inch of mulch in containers.

Growing Tips

Growing blue butterfly flower is no more complicated than establishing a watering routine and sticking to it. Fertilize monthly from the spring to fall, preferably with a complete all-purpose formula that includes both major and minor elements as well as trace minerals. When the mulch decomposes, work it into the soil and replace it promptly. Deadhead flowers and prune erratic stems to maintain an upright, rounded form.

Care and Propagation

Few pests bother blue butterfly flower, but watch for mealybugs and spider mites in dry seasons. Expand your collection by propagating from semi-hardwood.

Companion Planting and Design

Grow a row of blue butterfly flowers in the middle ranks of beds and borders or plant them in a blue pot as a centerpiece. For cool blue and white look, add star jasmines and a lily of the valley tree.

Try These

It has different leaves and flowers, but for something equally easy to grow, try flaming glorybower *Clerodendrum splendens*.

Those who admire Mexican hydrangeas, harlequin flowers, and flaming glorybowers or bleeding heart vines may not recognize their relatively quiet cousin, blue butterfly flower. Were it not for the others' corrugated leaves and spectacular flowers, this African native would no doubt get more attention or at least a more memorable name. Its leaves are smaller, smooth, and solid green, covering a plant dwarfed by its relatives. Their distinctive, gregarious flowers can be recognized from the balcony, while blue butterfly flower must be inspected up close to appreciate its subtle wonders. Scores of blue and white flowers dot the plant and oddly resemble butterflies hovering over the leaves; they are complex blossoms designed to attract its pollinators. Their intricate design and whimsical attitude combine with months of flowers and a neat growing habit to make blue butterfly flower a favorite tropical plant, grown as an annual in beds or at the center of a mixed container elsewhere.

Plant Family
Verbenaceae

Other Common Name
Blue glorybower

Bloom Period and Seasonal Color
Blue and white flowers appear in the summer and fall

Mature Height × Spread
Up to 10 ft. × 10 ft. where hardy; averages 3 ft. × 1 ft.

Blunt-leaf Peperomia

Peperomia obtusifolia

In its native haunts from Mexico to the West Indies, blunt-leaf peperomia is a robust forest denizen. It creeps stealthily over stumps and roots in leaf litter to form loose, low mounds dense enough to completely shade the ground below. The leaves are round and deep jade green, often found with random cream, yellow, or pink patterns in selected cultivars. Both stems and leaves are fat and waxy, an indication of their somewhat succulent nature. Their effect is lush yet stocky and muscular, as if the plants might overrun the onlooker. The plants bloom on upright, fuzzy spikes, tiny rows of green flowers packed tight like antennas over the leaves that are followed by oddly smooth fruits. Four of the peperomias native to Florida, including this one, are endangered species and should not be removed from the wild. Fortunately, there are plenty in cultivation to go around.

Plant Family
Piperaceae

Other Common Name
Baby rubber plant

Bloom Period and Seasonal Color
Dark green leaves, often variegated

Mature Height × Spread
Up to 1 ft. × 1 ft.

When, Where, and How to Plant

Blunt-leaf peperomia can be a groundcover in the tropics and there is no better container plant anywhere. Pots can spend their summers sunk into mulch beds in shade or dappled shade along with other compatible plants and move seamlessly to the indoor garden in the winter. Shade or partial shade keeps this plant deep green and growing; avoid direct sun, indoors or outside. Prepare an organic, well-drained soil by amending garden soil and potting mixes with organic matter if needed. Grow one in a standard 8-inch clay pot until it becomes overcrowded, or space plants 8 to 12 inches apart in beds. Mulch plantings lightly in beds and not at all in pots. If garden soils stay saturated, grow peperomias in pots regardless of your zone.

Growing Tips

Overwatering can cause this plant to rot at its roots and stems wherever it is grown. Water well and then allow the soil to dry out before watering again. The soil will look and feel dry and pots will be lighter when it is time to water again. Fertilize blunt-leaf peperomias lightly throughout the year and take care to keep granular products off the leaves. Repot annually to accommodate growing roots. Cut back beds of peperomias in late winter to stimulate new growth.

Care and Propagation

Watch for mealybugs and signs of fungus. Propagate by division and by rooting tip and leaf cuttings.

Companion Planting and Design

Create a classic tropical green scene with peperomias, dwarf scheffleras, jade plants, and night-blooming cereus. Add it to a group planting with bamboo palms, bush lilies, and crotons for color and contrasting plant textures.

Try These

'Gold Tip' has mottled variegation at its leaf edges and 'Minima' is a lovely smaller form.

Brazilian Red Cloak
Megaskepasma erythrochlamys

When, Where, and How to Plant
Brazilian red cloak can grow in sun, part sun, or part shade, as long as the soil is kept moist. It will develop a looser form and fewer flowers in the shade, but the plants will require less water too. Prepare a soil in garden beds or containers that is richly organic, fertile, and well drained. Amend as needed with compost and ground bark to meet these needs. Space young plants 2 feet apart in garden beds, plant three to form a large triangular clump, or grow one in a standard 12-inch pot. These plants flourish in garden soil in warm weather, root easily, and can be dug up in the fall and potted to spend the winter indoors in the temperate zone. Mulch plantings and containers with 1 to 2 inches of organic material.

Growing Tips
Water and fertilize these plants regularly and let their soils dry out only slightly between irrigations. Use a general-purpose garden fertilizer, either soluble or granular, as often as directed on the label from the spring through fall. Fertilize less often in the winter and when starting young plants. As mulch decomposes, work it in to the soil and replace it promptly with fresh material. Cut flowers for arrangements or deadhead them when they fade.

Care and Propagation
Few pests bother Brazilian red cloak but watch out for scale insects. Propagate by cuttings. See "Propagation" (page 36) for more details.

Companion Planting and Design
Grow this plant with copperleafs and chenille plants for colorful contrasts, or let it fill around red Abyssinian bananas and bird of paradise plants. Good companions include Persian shield and cat's whiskers.

Try These
Justicia brandegeana, the shrimp plant, offers a different take on the waffle-bract flower arrangement.

Brazilian red cloak sounds as if it might be a plant that drapes its leaves or wraps its stems but the common name refers to the hidden treasure of torchlike flowers. The plant's shape is that of a layered cake with flowers bright as candles on every tier. Great waffled spikes of red bracts protect two-lipped white flowers that finally poke out to create a tower of color. Their shape and arrangement resemble lollipop flowers, Pachystachys lutea, but red cloak's brilliant color and size take the effect to a higher level. The plants, too, elevate the image of evergreen leaves to include yellowish new growth and sharply drawn midribs that give each leaf the illusion of movement even in still air. In frost-free garden beds and containers everywhere, Brazilian red cloak's surprises include the long life of its flowers and the ecstasy it brings to butterflies.

Plant Family
Acanthaceae

Other Common Name
Red justicia

Bloom Period and Seasonal Color
Red bracts and white flowers sporadically through the year

Mature Height × Spread
Up to 10 ft. × 6 ft.

Bush Lily
Clivia miniata

Wildly popular, with hordes of fans in countries around the world, bush lilies are strangely ignored in the United States. There should be a covey of the South African natives in every atrium and office lobby to thrill visitors with their robust evergreen leaves. Every balcony and bedroom should be adorned with this lily's husky form and elaborate circles of flowers. The leaves are thick, dark green, and decidedly masculine in their effect, strappy like their relatives in the Amaryllis family, but thicker and in a single plane. The flower scape rises from ground level to the top of the leaves and opens into halos, or umbels, of funnel-shaped flowers. Most of the flowers are a creamy orange sherbet shade with sunny yellow centers that last for weeks. Attractive seedpods form soon and often produce viable seed if left on the plant to mature.

Plant Family
Amaryllidaceae

Other Common Name
Clivia

Bloom Period and Seasonal Color
Flowers in the winter and spring, usually orange with yellow centers

Mature Height × Spread
Up to 3 ft. × 1 ft.

When, Where, and How to Plant
A classic potted plant, bush lily can spend summers sunk into mulch beds in shade or dappled shade along with other compatible plants. They can also be planted directly into those same beds for the warm months, dug up in the fall, and repotted or stored like caladiums. Bush lilies grow slightly faster with bright light but they cannot tolerate direct sun, indoors or outside. In either setting, provide an organic, well-drained soil by amending garden soil and potting mixes with compost or ground bark. Plant one in a standard 10-inch pot or space tubers 8 to 12 inches apart in rows and beds. Mulch plantings lightly in beds and not at all in pots. If garden soils are likely to stay saturated, grow in pots.

Growing Tips
Overwatering can cause tubers to rot. Allow pots or plantings to dry out between waterings. Dig up flooded bulbs, let them air dry, and dust with sulfur before replanting. Fertilize monthly unless the tubers are in storage. Use a soluble, general-purpose formula as directed on the product label in the spring and summer, then at half strength in the fall and winter. Bush lilies will produce more leaves and blooms if you repot them only when their roots are very constricted.

Care and Propagation
Watch for mealybugs and spider mites. Propagate bush lilies from offsets or seed in the spring and summer.

Companion Planting and Design
Grow this lily with earth star, dumb canes, and false aralia in pots, or plant it with weeping figs and blunt-leaf peperomias for at least three shades of green in one vignette.

Try These
C. caulescens is long stemmed; hybridized with *C. miniata*, it produces a tall landscape plant.

When, Where, and How to Plant

Except for a sun-tolerant few, caladiums belong in shady spaces and those with filtered light. Caladiums will rot in cold soils, even the dwarf varieties often seen in mixed pots. Start the bulbs indoors in pots six weeks before transplanting to the garden or outdoor planters in the spring, or purchase young plants that are already leafed out. Prepare a soil by amending garden beds and potting mixes with compost and ground bark to make certain they are organic and fertile, yet well drained. Plant caladiums 8 inches apart in groups or one in a 6-inch pot.

Growing Tips

Water often enough to keep the soil moist but never flooded where soils are naturally heavy with clay. Fertilize regularly with an all-purpose formula. Maintain 1 to 2 inches of organic mulch around caladiums to prevent drying out between waterings. Cut off the flower spikes as soon as they emerge and apply ½ cup of compost around each plant for added nitrogen to stimulate new leaves. Outside the tropics, let caladium plants dry out before lifting the tubers for winter storage in a cool, dry room.

Care and Propagation

Watch out for slugs and snails on young plants. Propagate by dividing the tubers.

Companion Planting and Design

Brighten the shade under any tree with a ring of caladiums, or line a shady path and let them compete with impatiens and orchid pansies for unsurpassed color in the shade.

Try These

Classics include white 'Candidum', pink 'Fannie Munson', and speckled 'Carolyn Wharton'.

Heart-shaped and sometimes rippled, caladium leaves dominate shady gardens in many decidedly untropical places. Like ugly ducklings, the rough, swollen "bulbs" grow up to be luscious swans in a colorful flock on the garden floor. They create a mood of sunny optimism, crisp and fresh no matter how high summer's heat and humidity. Shades of red, pink, white, and green in endless combinations spew out as if they were designed in the spin-art booth at a school carnival. Caladiums travel well and thus are among the most popular tropical plants in temperate climes. First collected by Europeans in Venezuela and Columbia, nearly all caladiums in the trade today come from Lake Placid, Florida. Caladium bulbs are actually tubers, like potatoes, and like them, can have many eyes, or growing points. Top-quality caladiums have the most eyes and will send up many more leaves than lower-rated choices.

Plant Family
Araceae

Other Common Name
None

Bloom Period and Seasonal Color
Multicolored leaves in shades of red, pink, green, and white from the spring through fall

Mature Height × Spread
Up to 2 ft. × 2 ft., but usually smaller

Calico Plant
Alternanthera brasiliana

The word calico *brings to mind tricolored cats and quaint fabrics, but it should also be part of your tropicals vocabulary. That cloth and those felines, as well as this plant, owe their name to the sixteenth-century Indian port town where Europeans first acquired bolts of printed cotton material. The lure of such faraway places comes home to gardens everywhere in calico plant. Its 1-inch-long, pointy leaves are a riot in bold shades of purple, red, and pink, sometimes with lighter or greener overtones. Endlessly amusing leaves are its most outstanding feature, but sweet little clover-like flowers appear in the summer. Calico plants are quite popular outside the tropics, too, but can befuddle gardeners who expect relatives to at least resemble one another. These plants have more of a Brady Bunch look to them, with many variations under one roof. For rich colors in sunny, damp spaces, this bunch is unbeatable.*

Plant Family
Amaranthaceae

Other Common Name
Joy weed, Brazilian red hot

Bloom Period and Seasonal Color
Year-round multicolored leaves in purples through reds to hot pink; white flowers in the summer

Mature Height × Spread
Up to 2 ft. × 2 ft. in beds and containers; dwarf varieties less than 1 ft. × 1 ft.

When, Where, and How to Plant
Prepare a well-drained, fertile soil that is rich in organic matter for calico plants in garden beds and containers. Add compost, ground bark, or another aged organic matter to native soils and potting mixes to meet these basic needs. Make plans for very regular watering, unless the soil stays moist. Leaf colors will be most outstanding in full sun in consistently moist soils, but calico plants tolerate part sun and dappled sun. Indoors, provide bright light and high humidity. Close spacing, 6 to 8 inches apart, creates a thick stand and helpfully shades the roots to prevent drying out. Plant dwarf types and mixed containers even more closely together. Water dry soils before planting.

Growing Tips
Pinch calico plants to keep them compact, and prune leggy plants as needed to stimulate new growth. Keep soils evenly moist, especially in hot weather, and maintain 2 inches of organic mulch around the plants. Fertilize with a complete, balanced formula monthly to stimulate new leaves and work old mulch into the soil as it decomposes. Clip off flowers for the vase and to stimulate new growth. Calico plants may become perennial in the subtropics.

Care and Propagation
Few pests bother this plant although occasionally spider mites or leaf spot may appear. Propagate by 4-inch tip cuttings anytime.

Companion Planting and Design
Calico plants can edge garden beds, be small hedges, or fill the space between upright and trailing plants in containers. Grow them in front of lobster claw or dwarf ylang ylang for drama.

Try These
Look for these: 'Party Time', 'Brazilian Red Hots', 'Purple Knight', and 'Royal Tapestry'.

Cape Aloe
Aloe ferox

When, Where, and How to Plant
Plant outdoors in warm weather or grow in containers year-round. Cape aloe stands up to lower humidity and requires good air circulation in humid climates. Choose a site in full sun or nearly so, on a slight slope if possible to assist with water and air drainage. Prepare a well-drained, fertile garden soil or container growing mix by amending as needed with sand and ground bark. Plant or position pots of Cape aloe 2 feet apart and nestle each plant securely into its soil to ensure its stability. Do not allow water to puddle around it. Repot when roots appear in the drain hole or if water sits on the top of the pot due to root crowding on the surface.

Growing Tips
Cape aloe is not the aloe vera used widely for burns and other irritations; do not confuse the two. This plant is drought tolerant and will rot at its base if overwatered, especially in the winter. Water only when the soil feels dry up to the first knuckle of your finger. Indoors, provide a cool room and even less water than usual in the winter. After spring flowering, offsets form and can destabilize a plant if they are not removed. Fertilize Cape aloe with a soluble complete formula or its granular equivalent, four times annually.

Care and Propagation
Keep an eye out for mealybugs. To propagate, pot up offsets or start seeds in the spring and summer.

Companion Planting and Design
Those designing for drought tolerance can combine Cape aloes with bougainvilleas for contrast between the heavy-looking succulent leaves and lighter-than-air bouquets of colorful flowers.

Try These
Aloe vera is the medicinal aloe that should be in every kitchen garden as a burn salve.

All aloes deserve consideration in tropical garden design for their thick, succulent leaves and stunning flowers that last for weeks. Cape aloe, named for its native South Africa home, is simply the fiercest of them all, and one of the few that can develop a tall stem over time. The leaves have spines along their edges, are fat and wide at their base, and then curve upward to sharp points. Sometimes the leaves are pure green, but they can also reflect blue and red tones, depending on season and sunlight. From the center of the plant, the stately flower stalks arise, called racemes. Hundreds of tubular blossoms join together in orange columns brighter than any neon sign. Their color is a message to the nectar feeders in the vicinity to make frequent visits during the long bloom season. They do, and when it is hummingbirds that come to call, the sight is breathtaking.

Plant Family
Liliaceae

Other Common Name
Ferocious aloe

Bloom Period and Seasonal Color
Dark orange flowers rise on tall compound racemes in late winter and spring

Mature Height × Spread
Stems can reach 6 ft. above a 3 ft. × 3 ft. rosette of leaves; even larger in tropical zones

Cape Daisy
Osteospermum ecklonis

Native to South Africa, like all members of this plant family, Cape daisy has a central disk flower that is dark blue surrounded by a halo of white ray florets. Even among members of this common plant family, these plants stand out, their stems slightly more erect than most, with flowers that seem to toss their petals about in a flirtatious pout. The leaves can look rather coarse, like a sneering lip, dark green, longer than they are wide, with a tendency to curl downward slightly. They are a fresh shade of gray-green, lighter underneath, and grow into a thick mat that spreads rapidly into a groundcover. If further proof of the plant's wayward attitude is needed, consider that the species is a parent of scores of named selections and hybrid Cape daisies in a rainbow of flower colors. Almost succulent and just a bit rowdy, Cape daisies are favorite summer annual in the temperate zones.

Plant Family
Asteraceae/Compositae

Other Common Name
Blue-eyed daisy

Bloom Period and Seasonal Color
White flowers in the summer, sporadically year-round

Mature Height × Spread
Up to 1 ft. × 3 ft.

When, Where, and How to Plant
Cape daisies vary in their ability to withstand cold temperatures, but none can tolerate wet soils. Choose a sunny site for this plant with soil that is organic, fertile, and very well drained. Amend existing garden soils and potting mixes as needed with compost and ground bark to ensure these needs are met. Purchase young plants in the spring or start seed six weeks before your area's last frost or first monsoon. Space plants 8 to 10 inches apart in mass plantings or grow one Cape daisy in an 8-inch container that is wider than it is deep. Keep new plantings well watered for the first few weeks when you are growing it as a warm-weather bedding plant. Mulch plantings with 1 inch of organic material.

Growing Tips
Cape daisies are most often grown as annuals where they are not hardy. Give them a regular routine that allows the plants to dry out between waterings. Fertilize regularly with a complete, all-purpose garden formula, either a soluble product or its granular equivalent. Well-established plants can be fertilized less often, but do not allow the leaves to pale. Deadhead flowers as they fade to extend the flowering season. Clip stems off as needed to keep the plants tidy and encourage new growth and flowers.

Care and Propagation
There are few pests to worry about. Propagate by division and seeds. See "Propagation" (page 36) for details.

Companion Planting and Design
Cape daisies create a white carpet around bold green foxtail agaves, spineless yuccas, and devil's backbone. Combine them with gazanias, hibiscus, and bottlebrush for a crazy quilt of colors.

Try These
'Lemon Symphony' covers dark green foliage with sunny yellow daisies that have bright purple centers.

Caricature Plant

Graptophyllum pictum

When, Where, and How to Plant

Provide a site in sun or part sun with a reliable water source nearby. Because the canes may look bare at their base, choose a spot where other plants can grow in front of caricature plant. Its leaves will be pale if the plants are drought stressed and the plants may stop growing. Prepare a soil for garden beds or containers that is organic, fertile, and well drained by amending existing soils with organic matter such as compost and ground bark. Space caricature plants 2 feet apart in the tropics, 8 to 12 inches apart in beds elsewhere or in large mixed pots. They thrive in humid, warm weather outdoors in garden beds and can go through the winter as potted plants in the temperate zone. Water well before mulching. Blanket the plantings with 1 to 2 inches of fresh organic mulch.

Growing Tips

Water and fertilize these plants often enough to allow them to dry out just slightly between irrigations. Young caricature plants need regular water but mature plants can develop some drought tolerance in beds or containers. Pinch young plants to encourage a bushy form. Use a general-purpose fertilizer regularly from the spring through fall. Choose a complete, balanced formula in soluble or granular form. Mulch outdoor plants well in the fall but do not allow them to become waterlogged.

Care and Propagation

Caricature plant suffers from few pests, but watch out for nematodes in sandy soils. Propagate it by taking cuttings if you want to expand your collection.

Companion Planting and Design

Caricature plants look best from across the garden in groups with contrasting leaves such as copperleafs, chenille plants, yellow bells, and yellow cestrums. Edge it with licorice plants or Persian shield.

Try These

A contrasting cousin, 'Tricolor', has green leaves with yellow ribs and centers.

Probably native to New Guinea, caricature plant could be just another beautiful tropical plant with colorful leaves. This plant is no run-of-the-mill bunch of painted leaves, but a full-bodied, piquant variation on that classic theme. Taller than croton and more intricate in its designs, caricature plant has green or maroon leaves that are thinner and more pointed than most, with a sultry curve along their edges. Leaf shades range from true green to chartreuse and chocolate, from cream to rosy pink and blood red. The center leaf ribs morph into creamy, pink, or light green blotches in a testament to genetic diversity, as varied as inkblot tests and just as unpredictable. A row of caricature plants nods when any breeze rustles the stems, top-heavy with vibrant leaf clusters. When bunches of small red flowers light up the scene, they become pompon-waving cheerleaders for the entire garden team.

Plant Family
Acanthaceae

Other Common Name
Café con leche, Jamaican croton

Bloom Period and Seasonal Color
Striking variegated leaves in shades of red and purple with small red flowers in the summer

Mature Height × Spread
Up to 6 ft. × 4 ft.; averages 4 ft. × 2 ft.

Carnation of India
Tabernaemontana divaricata

Like a spy with six passports, this plant readily changes its name, but none reveals all its secrets. The plant called carnation of India is native to India, but also to Myanmar and Thailand. Its flowers have the sweet clove fragrance of carnations but look papery, so it is also called crepe jasmine. The evergreen shrub or small tree is not a jasmine and has leaves reminiscent of gardenia (yet none of its names indicates that). The flowers can be completely double like a carnation or gardenia or they form pinwheel shapes, and thus the plant is called pinwheel jasmine. Perhaps the subterfuge serves to enhance the exotic image of this sturdy, robust plant with a wide range of comfort zones. At home in the tropics but hardy in the subtropics and sometimes slightly farther north, carnation of India is easily rooted for sharing with friends or adding to your indoor garden.

Plant Family
Apocynaceae

Other Common Name
Pinwheel jasmine, crepe jasmine

Bloom Period and Seasonal Color
White flowers in the spring and summer

Mature Height × Spread
Up to 6 ft. × 4 ft.

When, Where, and How to Plant
Carnation of India grows well in sun. In part sun, it will need slightly less water but will bloom less abundantly also. Prepare a fertile, organic, well-drained soil by improving those qualities in existing garden soil and potting mixes. Use compost, ground bark, and other aged products such as manure as soil amendments. These shrubs require consistent moisture and cannot tolerate flooding or drought without damage. Provide a wall or other protection from strong winds and salt sprays for carnation of India. Space young plants 2 feet apart in beds or mixed planters, or provide a standard 10-inch pot for a pair of rooted cuttings. Make sure to plant at the same depth they were growing originally, and water new plantings very regularly.

Growing Tips
Maintain consistent soil moisture. Use drips, soakers, or another reliable irrigation method that will direct ample water to the base of the shrubs. Mulch plants to a depth of 2 inches in beds, 1 inch in pots. Use a complete granular formula in the spring and summer; as organic mulch rots naturally, work it into the top layer of the soil in pots or beds. Prune lightly during the growing season to shape the plants and to stimulate new growth.

Care and Propagation
There are few pests. Propagate by cuttings taken from semi-hardwood in summer and by seed.

Companion Planting and Design
Plant a white garden featuring carnation of India with star jasmines, white angel trumpets, and moonflowers. Feature plants with shiny green leaves with contrasting textures by combining sago palms, yellow bells, and ruffle palms.

Try These
'Flora' has white, completely double flowers.

Cat's Whiskers

Orthosiphon aristatus

When, Where, and How to Plant

Cat's whiskers plants flourish in garden soil in warm weather, root readily, and are easy to maintain as potted plants over the winter where necessary. Pots can also be sunk into mulch beds outdoors in warm weather and returned to the indoor garden in late fall. Choose a site in full sun or mostly sun with a reliable water supply nearby. Consider using soaker hoses or another irrigation method that can be automated. Prepare soil for beds or pots that is well drained and rich in organic matter. Use compost or ground bark to amend existing soils if needed. Space plants 8 inches apart in garden beds and mixed planters, or plant three small plants in a standard 12-inch pot. Mulch beds and pots with 1 to 2 inches of organic material.

Growing Tips

Water cat's whiskers regularly and let the soil dry out only slightly between irrigations. Use a general-purpose garden fertilizer, either a soluble or a granular product. Apply it as often as directed from the spring through summer and less often in the winter. As mulch rots, work it in to the soil around the plants and replace it promptly with fresh organic material. Pinch or prune the plants to encourage thick growth and more flowers.

Care and Propagation

Watch out for slugs and snails. If you want more plants, propagate by taking cuttings.

Companion Planting and Design

Fill a space with low growers including cat's whiskers, summer snapdragons, coleus, yellow shrimp plants, and Persian shield. Use it to edge plantings with vines like golden trumpets, Mexican flame vines, and moonflowers.

Try These

O. labiatum, pink sage, is another member of the mint family that attracts hummingbirds; it has pink blooms.

For every gardener who tires of waiting for a reluctant plant to get growing and bloom, cat's whiskers can quickly become a keeper. Its fast growth, tidy habit, and perky personality easily make it an ever-present part of the garden scene. White or purplish flower spikes pop up suddenly and burst into clever lipped flowers with the long, curvy stamens that give the plant its most popular common name. Native to Southeast Asia and Australia, cat's whiskers fills sunny and shady spaces with dense, irregular shapes and flowers at every level. The result is a pleasant riot that begs not to be tamed by pruning, with flowers placed randomly in a cockeyed version of a birthday cake's candles. Cat's whiskers is a pleaser of a plant, eager to grow and bloom with little attention from its gardener beyond simple routine maintenance. It's no wonder so many people keep cat's whiskers, root it, and pass it along.

Plant Family
Lamiaceae

Other Common Name
Java tea

Bloom Period and Seasonal Color
White flowers from the spring through fall

Mature Height × Spread
Up to 5 ft. × 3 ft. where hardy; averages 2 ft. × 2 ft.

Chenille Plant
Acalypha hispida

Chenille patterns look like fuzzy caterpillars crawling across bathrobes and bedspreads, proudly raised in swirls. The chenille plant replicates that brushy, soft look with hundreds of tiny flowers gathered into chains as big around as a finger and up to 2 feet long. Their tight arrangement is unusual and eye-catching in the extreme. The classic red-purple clusters bloom for weeks, draping like a curtain thrown over the big, medium green plants. Bushy and dense with wide oval leaves, chenille plants are thick enough to form a hedge or fill large containers. They can reach great heights in their native Malaysia and other frost-free environments but respond well to pruning to a more manageable size in the garden and in pots. As the word chenille fades from use, it may be best to rename this one. An equally descriptive name could be dreadlock plant, since the thick tails of flowers also resemble this popular hairstyle.

Plant Family
Euphorbiaceae

Other Common Name
Red hot cattail

Bloom Period and Seasonal Color
Red, magenta, and white flowering chenille plants bloom year-round in frost-free areas and in pots, in the summer and fall elsewhere

Mature Height × Spread
Up to 10 ft. × 5 ft. in the tropics; up to 4 ft. × 2 ft. as an annual and in pots

When, Where, and How to Plant
Chenille plant is easily grown from seed, and plants are available in garden centers in the spring. Start seeds in peat pots indoors about six weeks before the last frost is expected, or anytime in frost-free areas, including indoor gardens. Prepare a soil that is well-drained yet organic and fertile for garden beds and containers by adding organic matters or sand to native soils and potting mixes. Provide a partly sunny location outdoors and plant in warm weather. Space chenille plants 12 to 18 inches apart in beds. Indoors, a bright indoor room provides plenty of light to bring on the flower show. Plant 2 or 3 chenille plants in a 10-inch hanging basket to show them off indoors or on a porch. Turn the pots monthly to ensure flowers develop evenly around a plant.

Growing Tips
To keep the well-drained soil moist and maintain nutrition levels for chenille plants, establish a rhythm for water and fertilizer applications. Let the soil dry slightly between waterings. Fertilize young plants very regularly. Deadhead the flower clusters before they fade completely to stimulate rebloom.

Care and Propagation
Watch for mealybugs, aphids, and spider mites on indoor plants. To propagate, take 4-inch tip cuttings from non-blooming branches or grow from seed. See "Propagation" (page 36) for details.

Companion Planting and Design
Go for a cottage garden style by pairing it with summer snapdragons, Philippine lilies, and cleome in beds.

Try These
Dwarf chenille plant (*A. pendula*) is excellent used as a groundcover or basket plant. *A. hispida* 'Alba' is white.

Chinese Hat Plant

Holmskioldia sanguinea

When, Where, and How to Plant

Chinese hat is a plant of moderation. It is not drought tolerant but does not tolerate wet feet either. Put the plants in sun or part sun with a reliable water source nearby. Prepare a well-drained, organic, fertile soil for beds and containers. Amend garden soils and potting mixes with sand and organic matter such as compost and ground bark to achieve those conditions. Space plants 2 to 3 feet apart in beds or planters or grow one in a container 12 inches deep and wide. Make room in garden beds outdoors year-round in the tropics or in warm weather elsewhere, then pot up Chinese hat plants for wintering indoors in a bright, warm room. Keep plantings mulched with 1 to 2 inches of organic material.

Growing Tips

Chinese hat plants need to be watered regularly and allowed to dry out only slightly between watering. If young plants sprawl, prop them up to keep leaves clean until they can stiffen with age. Use a general-purpose fertilizer regularly, preferably a complete formula with trace elements included. As mulch decomposes, work it in to the soil and replace with fresh organic material. Remove the flowers to keep the leaves growing.

Care and Propagation

Watch for whiteflies and scale insects. If you like, propagate by tip cuttings.

Companion Planting and Design

Grow Chinese hat plant near its lookalike, bougainvillea, for an explosion of bracts. Combine for strong contrast with yellow cestrums, polka dot plants, Persian shield, and fan flowers.

Try These

Tahitian hat plant has lavender and white flowers, while other selections offer yellow- and salmon-colored flowers.

Arching, sprawling, and blooming all over, Chinese hat plant waltzes into the garden party with unstoppable colorful élan. Blazing hot flowers crowd the leaf axils on every branch, their orange sepals like Frisbees surrounding stunning scarlet red tubes. The stamens stand up from their sepal bed, held tightly in some plants and splayed open in others to reveal dancing butterfly shapes. Chinese hat plant is native to the foothills of the Himalayas and the Malay Peninsula and is grown outdoors everywhere in frost-free zones. The leaves are light green with dark veins and shaped rather like those of althea. The plants can be at home almost anywhere as a bulky, fountain shape perfect to grow alone or spill over a wall. Its positive reaction to pruning enables a ready adaptation to espalier or column supports that will bring the flowers up to eye level.

Plant Family
Lamiaceae

Other Common Name
Cup and saucer plant

Bloom Period and Seasonal Color
Red and orange-red flowers in the summer and fall

Mature Height × Spread
Up to 6 ft. × 6 ft.

Coleus

Coleus × hybridus

The coleus plants grown in twenty-first-century gardens are virtually all hybrids, and thus have no native home in the wild. But in one genetic way or another, they are the offspring of Solenostemon scutellarioides, *native to Southeast Asia and the Malaysian peninsula and noted for their outrageous leaves. This plant is a childhood memory for many, since that venerable coleus was a staple of 1950s and 1960s gardens. Like many of its relatives in the Mint family, coleus is an herbaceous, or green-stemmed, branching plant. Like some of them, its flowers are inconsequential and even undesirable compared to its leaves. And oh, those leaves! They are green, yellow, red, purple, cream, and orange in endless combinations from splattered mosaics to neatly edged two-tones. Coleus offers a plethora of round, oval, pointy, ruffled, cupped, curled, and serrated leaf shapes. For bizarre, leafy, tropical excitement, nothing beats a modern coleus.*

Plant Family
Lamiaceae

Other Common Name
Painted nettle

Bloom Period and Seasonal Color
Multicolored, multipatterned leaves all year

Mature Height × Spread
Up to 2 ft. × 4 ft.; averages 1 ft. × 2 ft.

When, Where, and How to Plant
Coleus is not drought tolerant nor can it tolerate flooded soils for more than a few hours. It thrives in any garden soil or potting mix that drains well and is rich in organic matter. Those conditions are best met by adding compost, ground bark, and enough sand to make the soil feel gritty to the touch. Select a site for coleus in sun, part sun, or shade, depending on the cultivar. Lay out soaker hoses in their bed or arrange another reliable irrigation method. Space plants 10 to 12 inches apart in rows or provide a standard 8-inch container for each plant. Water new plantings well and mulch with 2 inches of organic mulch such as ground bark. Pinch the tips out of each stem to begin the branching process.

Growing Tips
With a simple routine of regular care, coleus plants are very easy to grow. Water often enough to prevent wilting and soak the plants deeply each time to encourage deep rooting and stability in stormy weather. Add a soluble all-purpose formula fertilizer to the water monthly. Mix at full strength in the spring and summer, then at half strength in the fall and winter. Deadhead flowers immediately and pinch plants back regularly to encourage bushy growth.

Care and Propagation
Watch for mealybugs and aphids. Coleus are easy to propagate; root tip cuttings in potting mix or water.

Companion Planting and Design
Use upright coleus with contrasting chenille plants, summer snapdragons, yellow shrimp plants, and angel trumpets. Create a summer coleus hedge with close spacing and regular pinching.

Try These
You are not limited to the shade anymore. Sun coleus varieties have been bred for their increased sun tolerance.

Copperleaf
Acalypha wilkesiana

When, Where, and How to Plant
Copperleaf, or copper plant, displays its most brilliant colors in sunny sites where it grows rapidly to produce a dense, leafy shrub. Choose a location for copperleaf that is away from wind and salt spray. Leaf patterns and colors are undaunted by reflected light and heat. That makes copperleaf an excellent choice near swimming pools and concrete paths or patios. Use one copperleaf or a group as a focal point in full sun surrounded by small ornamental grasses, or to create a more subtle effect as a background for caladiums in shady spots.

Growing Tips
Copperleaf flowers should be removed if they appear on young plants. Regular use of a nitrogen fertilizer, such as cottonseed meal, will encourage leaf growth over flower development. Although they grow best in well-drained soils, copperleaf plants suffer if they are allowed to dry out completely between waterings. Wilted copperleafs are difficult to rehydrate and may suffer leaf drop.

Care and Propagation
Be on the lookout for mealybugs and mildew. To propagate, take 4-inch tip cuttings from semi-hardwood and use a heating mat for fastest rooting.

Companion Planting and Design
Companion plants can include sanchezias, fatsias, and others that tolerate consistently moist soils.

Try These
'Macrophylla' has bronze, yellow, cream, and red variegation. 'Musaica' looks mottled in orange and red tones. Named selections include pink 'Showtime' and 'Beyond Paradise'.

Adding powerful contrast is essential in a world full of solid green leaves, which makes copperleaf a natural addition to garden designs and combination planters. These robust, bushy shrubs offer stunning contrast, covered as they are in red, yellow, copper, mahogany, and even pink-tinged leaves. Some copperleaf varieties have leaves that are narrow and drooping, others that are pointed and wide in the middle, and some have curls and crimps. Native to Fiji and other islands in the South Pacific, this Acalypha species has given rise to more than two dozen named varieties and selections. Each is prized for its particular leaf patterns and color combinations seldom seen in other plants. No two leaves are exactly alike, but all are slightly serrated, mostly oval, and arranged alternately along dark stems. Outdoors, copperleaf plants can produce a 4-foot-tall hedge in just a few months. Potted copperleafs retain their color indoors in a bright, humid environment.

Plant Family
Euphorbiaceae

Other Common Name
Copper plant

Bloom Period and Seasonal Color
Prized for leaf colors including bronze, copper, red, and gold

Mature Height × Spread
Up to 10 ft. × 10 ft. in the tropics; averages 4 ft. × 4 ft. in the subtropics; to 3 ft. in pots

Coral Plant
Jatropha multifida

Just when it might seem that every possible arrangement of flower parts has been found in tropical plants, coral plant takes the cake for surreal beauty. A small, shrubby tree, it tosses an open crown of big, gently drooping, dark green leaves atop thick, succulent trunks. Each leaf is white underneath and cut into nearly a dozen deep lobes, long, narrow, and curiously pointed down as if there were curtain weights in its tip. Standing high above the leaves on purplish stems, coral plant's flowers personify the term weird. A plethora of star-shaped flowers is gathered into stiff, neon pink, or candy red clusters that do look strangely like clumps of coral. They are almost rigidly erect and seem separated in space from the leaf skirts below. As the flowers mature on this incongruously irresistible plant, they produce olive green and olive-shaped seedpods that complete a plant worthy of Dr. Seuss's garden.

Plant Family
Euphorbiaceae

Other Common Name
Guatamala rhubarb

Bloom Period and Seasonal Color
Scarlet red and coral flowers from the spring through fall

Mature Height × Spread
Up to 20 ft. × 10 ft.; averages 8 ft. × 4 ft.

When, Where, and How to Plant
Young coral plants will thrive in a sunny or mostly sunny site with a reliable water source nearby. The plants are adaptable to many soils but thrive in garden beds or containers where the soil is fertile and has very good drainage. Amend existing garden soils and potting mixes with sand and ground bark to achieve these conditions. Space plants 12 inches apart in garden beds and mixed planters, or plant two small coral plants in a standard 12-inch pot. Mulch beds with no more than 1 inch of organic material. Do *not* mulch pots, unless it is with gravel. Like most *Euphorbia* species, coral plant sap is toxic and can be a skin irritant. Wear gloves when handling the plants.

Growing Tips
Coral plants respond well to routine maintenance that enables the soil to dry out just slightly between irrigations. Water the plants less in the winter so they rest and expect some natural leaf drop. Those leaves will be replaced when growth resumes. Use a general-purpose garden fertilizer as directed from the spring through fall, but less in the winter (if at all). As mulch decomposes naturally, work it in to the soil and replace with fresh mulch promptly.

Care and Propagation
Another tropical not bothered much by pests. You can propagate by cuttings in the spring or from seeds.

Companion Planting and Design
Grow coral plants in a succulent garden with rose cactus or surround them with silver vase bromeliads for striking leaf contrasts. Let them shine with the darker greens of sago palms and curry leaf.

Try These
Peregrina (*J. integerrima*) has heart-shaped leaves and more traditional star-shaped flower clusters.

Croton
Codiaeum variegatum pictum

When, Where, and How to Plant
Choose named croton varieties with improved color and leaf retention in low light. Indoors, provide strong light and a spot away from drafts and heating vents. Outdoors, find a site in part shade or dappled sun. Grow crotons in raised beds or in containers sunk into mulch beds during warm weather. Leaving them in containers eases the transition indoors where temperatures fall below 50 degrees F, but the plants can also be lifted from their beds and potted in the fall. In either case, prepare a soil that drains well and is richly organic. Crotons are not drought tolerant, nor can they stand wet feet or flooded conditions. Mulch raised bed plantings with 1 inch of organic material such as ground bark. Do not mulch potted crotons.

Growing Tips
Overwatering can cause crotons to drop their leaves and rot at the roots. Always let pots or plantings dry out slightly between watering. Fertilize four times annually with an all-purpose formula, preferably one with major and minor elements and trace minerals included. Unlike most tropical plants, crotons prefer relatively cool temperatures. Mulching around the plants outside and spacing to allow air circulation around pots will cool the microclimate.

Care and Propagation
Watch out for mealybugs and keep leaves free of dust. Propagate by dividing clumps or from stem cuttings.

Companion Planting and Design
Let crotons be the colorful contrast to dwarf schefflera, dumb canes, false aralia, and bush lilies. Make a mulch bed off a deck and sink a dozen croton plants into it for high drama.

Try These
'Petra' has thick, mostly yellow leaves and tolerates lower light, as does 'Norma'.

Crotons could be called the "paradox plant" because they know how to keep their raucous leaf colors a secret until just the right moment to dazzle onlookers. Leaves start out green and soon develop incredibly diverse patterns in colors so rich they might be painted in oil. Croton is native in a huge swath of the South Seas from Australia to Java, where they developed myriad leaf shapes and color mixtures. Their shapes run the gamut from large and broad, almost round, to small, from oak leaf to spiral and narrow leaf types with limitless variations on each theme. From green to cream, red, orange, and nearly pink, the leaves are thick and waxy enough to warrant a double take, just to make sure they are not artificial. Like their relatives poinsettia and crown of thorns, crotons leak a white sap when punctured that may discolor clothing and irritate skin.

Plant Family
Euphorbiaceae

Other Common Name
Rushfoil

Bloom Period and Seasonal Color
Prized for its leaf colors including yellow, cream, green, and many shades of red

Mature Height × Spread
2 ft. to 4 ft. × 1 ft. to 2 ft.

Crown of Thorns
Euphorbia milii

Small and spiny, crown of thorns swaggers into the garden like a wrestler. Sometimes standing awaiting any attack, sometimes crawling on the floor of the ring ready to strike, the plants capture all eyes whether they are in bloom or not. Native to Madagascar, its gray-brown stems are covered in wicked spines, creating a frankly defensive posture. The leaves are spade-shaped, 1½ inches long, and sparkling emerald green; as individuals they are short-lived in dry environments. At times a plant is nearly leafless, but leafs out again when rain returns in the wild or once watering is resumed. The unexpected flowers give the plant its name, recalling the crown of the crucifixion. Solid colored bracts open wide to attract pollinators to the tiny flowers within. Crown of thorns covers every branch in bright flowers that wait for those brave enough to face the spines and reach the precious, nectar-filled blooms.

Plant Family
Euphorbiaceae

Other Common Name
Christ-thorn

Bloom Period and Seasonal Color
Red, salmon, orange, yellow, or white bracts and greenish yellow flowers bloom sporadically through the year

Mature Height × Spread
Up to 4 ft. × 1 ft.

When, Where, and How to Plant
Crown of thorns grows and flowers best in very well-drained, even sandy soil. Where this condition cannot be met, pot it up in a very well-drained growing mix. Bagged potting mixes intended to maintain moisture and fertility may provide conditions that are too wet and which promote leafy growth (over blooms) unless it is well amended with sand and finely ground bark. Grow crown of thorns in a pot 2 inches wider all around than its stem or combine it in a mixed container with other plants best grown on the dry side. Choose a site in sun or mostly sun, and provide ample space for good air circulation around each plant. The plants will tolerate moderate amounts of salt spray and persistent breezes. Protect the pots in the rainy season and when temperatures drop below 50 degrees F.

Growing Tips
Allow crown of thorns plants to dry out thoroughly between waterings. The plants are drought tolerant but when they are flowering they do need water. When the buds appear, water the plants regularly but not excessively. Delay repotting until the pots tip over or break. Fertilize crown of thorns regularly but sparingly. Use a soluble flower formula every other month or at half strength monthly. Move pots into a sunny, warm room to overwinter outside of dry, frost-free areas.

Care and Propagation
Luckily for the gardener, it suffers from few pests. Propagate by cuttings and seed. See "Propagation" (page 36) for details.

Companion Planting and Design
By the pool and other hot spots, let crown of thorns and desert rose take center stage. Grow some with several shades of bougainvilleas for a stunning display of hardworking flower bracts.

Try These
'Hislopii' has tall spikes of coral flowers, while 'Lutea' has sweet yellow flowers.

Cut-Leaf Philodendron

Philodendron bipinnatifidum

When, Where, and How to Plant

Cut-leaf is not as rough as it looks and will take on a bronzy tone in full sun. Select a site in bright shade, indoors or out. Plants can grow outdoors as a summer annual plant in garden beds and are overwintered easily outside the tropics and warm subtropics. Prepare a soil that is fertile, organic, and well drained. Most garden soils and potting mixes will benefit from amendments of compost or other organic matters to improve their condition. Space cut-leafs 3 feet apart in beds, or provide a standard 10-inch pot for one young plant. Water new plantings well, and consider a soaker hose for these water lovers. Mulch all plantings with 2 inches of organic matter and, as the mulch rots, work it into the soil and replace it promptly.

Growing Tips

Cut-leaf philodendrons need ample amounts of water regularly applied to grow at their best. Do not let the soil dry out. If bronzy patches appear on the leaves, move the plant into less light. Fertilize regularly with a soluble formula mixed into their water, or a granular slow-release formula. Dig up garden plants or propagate them in late summer for wintering indoors where necessary. Because they are hardy into the warm subtropics, they are good candidates for cool storage strategies too.

Care and Propagation

Watch for mealybugs and aphids. Propagate by cuttings that appear as shoots around the base of plants.

Companion Planting and Design

Cut-leaf philodendrons create great green drama along with dumb canes, nun's orchids, lady palms, and Malaysian orchids. Use them as a centerpiece edged with chocolate plants and peacock plants or caladiums and impatiens.

Try These

'Variegatum' has delicious yellow leaf markings and many hybrids have cut-leaf as a parent.

Nothing looks like cut-leaf philodendron, long known as selloum but renamed to better describe its leaves. It is a staple of the tropical garden and widely grown in large interior spaces. The deeply cut, bipinnate leaves are exuberant, with stiff midribs that hoist the huge leaves into full, glorious view. They are a shiny, almost bright green on their upper surfaces and quite drab on the undersides. Two other names for this plant are tree philodendron and split leaf philodendron, but most are called cut-leaf, particularly the many hybrids that have this plant as a parent. "Split leaf" is descriptive, since the leaves are split into many leaflets, and "tree" refers to the fact that, unlike most of its kin, this plant does not climb readily. Over time, this native of Brazil and Paraguay will climb in the tropics, and everywhere it will develop a woody stem with ever more complex leaves that will split again and again.

Plant Family
Araceae

Other Common Name
Selloum, tree philodendron, split leaf philodendron

Bloom Period and Seasonal Color
Green leaves in all hues, depending on available sunlight; white flowers in the spring

Mature Height × Spread
Up to 15 ft. × 15 ft.; averages 3 ft. × 3 ft.

Desert Rose
Adenium obesum

A few plants can turn anybody into a gardener, and desert rose is one. The flowers are perky and long-lasting tubes. Yellow throats open into five fat lobes, colored in solid rich colors or with brightly marked scalloped edges. Desert rose has medium green leaves that are cheerful, boxy, and thick—a perfect backdrop for such bold flowers. The beauty of desert rose is matched by its low-maintenance reliability. Full sun, even inside a west-facing window, does not wilt desert rose, and forgetful folks appreciate its drought tolerance. Its attractive, gray stem is a swollen caudex, a sturdy stem that stores water very efficiently and gives this succulent a treelike look even when young. A native of South Africa and Madagascar, desert rose is a long-lived, evergreen plant that is untroubled by dry environments, indoors or out. For full-on tropical effect with virtually no maintenance, nothing surpasses desert rose.

Plant Family
Apocynaceae

Other Common Name
Impala lily

Bloom Period and Seasonal Color
Spring and fall primarily, solid and bicolor flowers in shades of red, pink, and white

Mature Height × Spread
Up to 5 ft. × 3 ft. in the wild; averages 2–3 ft. × 1–2 ft. in planters.

When, Where, and How to Plant
Desert roses are best grown as container plants outside the tropics. They grow slowly but steadily in tiny bonsai trays or classic 6- to 10-inch clay pots or combined in huge concrete planters with other plants that require very little water. They are suited to sunny sites outdoors in hot summer weather. But desert rose ranks among the very best plants for indoor gardening all year in rooms too sunny or hot for most other plants. Prepare a very well-drained soil such as a lightweight potting mix/sand combination. When transplanting, take care to keep the base at the same level it was growing before. A sunken base may get fungus disease that can stunt growth or even be fatal. Repot as needed to maintain 2 inches between the caudex and the pot's edge. To maintain its size, root prune and repot in same container.

Growing Tips
Easy to grow in dry, hot, sunny conditions, desert rose can be pruned in early spring while it is young to stimulate more branches that will bear flowers. Additional pruning may be needed to keep a bushy shape and control height on mature plants. Water desert rose only when it is dry, which will be more often in the summer than the winter. Fertilize once or twice monthly in the spring and summer by watering with a complete soluble fertilizer mixed at half strength.

Care and Propagation
Watch for mealybugs and sometimes spider mites. To propagate, take 3- to 4-inch tip cuttings and let them "heal" for a day before rooting.

Companion Planting and Design
Group pots of sun-loving plants with good drought tolerance on your sunniest windowsills, decks, and patios. Combine desert rose with Cape aloes, bougainvilleas, and chameleon plants.

Try These
'Taiwan Beauty' has pink flowers with rosy throats. 'Uranus' has white flowers with bright pink edges.

Devil's Backbone

Euphorbia tithymaloides

When, Where, and How to Plant

Select a sunny location for garden beds or pots spending the summer outdoors, either sunk into mulch beds or planted out during warm weather. Indoors, a cool, brightly lit room will sustain the plants over the winter with minimal watering. Amend existing garden soils and potting mixes with coarse sand and ground bark to grow devil's backbone. The soil must be well drained and fertile, but with a neutral pH, so hardwood bark is preferred. Grow two small plants in a standard 8-inch clay pot, or space them 1 foot apart in beds and mixed plantings. The stems can be fragile but any broken during transplant will root readily. Mulch lightly, with no more than 1 inch of organic matter, to prevent weeds around this upright plant.

Growing Tips

Water devil's backbone well and wait for it to dry out before watering again. Overwatering will cause the weak stems and a floppy appearance that could be mistaken for wilting. Fertilize regularly but lightly, mixing a general-purpose soluble formula at half the recommended strength. Repot annually or as needed to accommodate the stems as they multiply. Cut back any stems that become weak after flowering.

Care and Propagation

There are few pests to bother it. Propagate by stem cuttings. See "Propagation" (page 36) for details.

Companion Planting and Design

Grow a powerful scene with devil's backbone, crown of thorns, wax plants, and coral plants. Soften the devilish impact with dwarf powder puffs, ti plants, hibiscus, and kanga paw in a mixed planting.

Try These

'Varigatus' has the most mottled leaf patterns, but even they range wide in pink-white-green mixes.

A plant with such an evil common name should not be so easy to love. The name connotes danger, yet devil's backbone has no thorns and is neither sparse nor rough-textured. Indeed, it is a wildly diverse, unearthly thing with stems that look as if they are covered in fondant that is green or blue or blue-black. As thick as pencils, they are sharply angular like rickrack and give the plant one of its names, the zigzag plant. The leaves are wavy and vary in the species and its most common forms, from solid green or shades of green to edging in white and pink. Star-shaped flower clusters grace the stems; usually laid out in one flat plane, they resemble propellers. Some say the flowers of devil's backbone look like cardinals perched around a feeder, each flower a long-tailed red teardrop that gives rise to another common name, red bird cactus.

Plant Family
Euphorbiaceae

Other Common Name
Zigzag plant, red bird cactus

Bloom Period and Seasonal Color
Green leaves often variegated with light green, cream, and pink

Mature Height × Spread
Up to 3 ft. × 3 ft.

Dumb Cane
Dieffenbachia spp.

If there were an Emmy for the actor with the longest-suffering role as a houseplant, dumb cane would win every year. Native to the American tropics, these hunky, buff plants thrive in warm, humid spaces where direct sunlight never falls yet it is their karma to toil in hot, dry, sunny places like office buildings and atriums. Most modern plants are descendants of an interspecies hybrid that can be remarkably easygoing and take years to succumb to the trials of dusty leaves and forgotten fertilizers. Named for pioneering naturalist Ernst Dieffenbach, the plants are trailblazers, too, often the first potted plant in a burgeoning collection. It may be their lot to suffer from the mistakes of novice gardeners, in part to pay back the many times their toxic sap has been used for evil. Suffice it to say that all parts of this plant are poisonous, presumably to deter prey from nibbling its lush, wildly patterned leaves.

Plant Family
Araceae

Other Common Name
Often called by cultivar names, such as 'Rudolph Roehrs' or 'Hatusko'

Bloom Period and Seasonal Color
Grown for evergreen patterned leaves in cream, white, yellow, and green shades with green spathes and white spadis outdoors in the spring

Mature Height × Spread
Up to 6 ft. × 3 ft.

When, Where, and How to Plant
Dumb canes thrive in bright light and filtered sunlight. Prepare the soil by amending garden beds and potting mixes with compost and ground bark to make certain they are organic and fertile, yet well drained. Space plants for good air circulation in areas where regular rainfall will keep the leaves wet for days. Otherwise, encourage humidity by grouping plants. A plant 2 inches in diameter needs a pot or space that is at least 6 inches in diameter, slightly more for a group planting. Transplant carefully to install plants at the same level they were growing originally. Transplant potted dumb canes into larger pots with fresh soil annually if continued growth is desired, or root prune and repot in same container to control a plant's size.

Growing Tips
Maintain very simple routine care: water, dry out just briefly, and water again. They are not drought tolerant, nor do they tolerate standing in water, and may be best grown in containers outside the tropics. Sink the pots into beds of mulch outdoors during the warm months and move them indoors when temperatures drop below 50 degrees F. Fertilize regularly to promote steady growth. Mix a soluble all-purpose formula into the water once monthly all year.

Care and Propagation
Watch for mealybugs and snails. Propagate by rooting canes and making air layers.

Companion Planting and Design
Combine with Persian shield or caladiums for cool pinks and greens. Grow some with angel wing begonias for complete contrast in form and color.

Try These
'Exotica Alba' has amazing, almost solid white leaves with green edges and 'Tropic Honey' is cream colored.

Dwarf Cone Ginger
Costus spicatus

When, Where, and How to Plant
Choose a site in part shade, dappled sun, or mostly sun for this adaptable plant. Be sure there is a reliable irrigation option in place and be aware that water demands increase with sunlight. Dwarf cone gingers need consistently available moisture as well as good drainage to grow and bloom. Prepare soil for containers or garden beds that is organic and fertile. Amend soils as necessary with organic matters to achieve these goals. Use a combination of materials, such as ground bark and compost, for best results. Plant gingers 1 foot apart in beds or one clump in a standard 10-inch container. Mulch garden plantings with 2 inches of organic material such as ground bark. Limit mulch in pots to 1 inch depth.

Growing Tips
Keep soils in beds and containers consistently moist but not flooded. Water this ginger often enough to maintain both its good green color and steady growth. Fertilize the plants with a complete balanced formula product, regularly but not often. Use granular products three times annually; mix soluble products with water every other month from the spring to fall. Reduce watering and fertilizer in the winter. As mulch decomposes, work it in to the top of the planting and replace it.

Care and Propagation
There are few pests to bother this plant. Propagate by dividing the clumps and by taking cuttings.

Companion Planting and Design
Combine dwarf red gingers with taro, sago palms, and Persian shield for vivid colors and contrasts. Plant a skirt of licorice plants or bat face cupheas around it in a big pot.

Try These
C. afer, spiral ginger, is taller and is hardy in the warm subtropics.

Dwarf cone ginger strides confidently into every garden design, a frankly macho plant. This one fills a garden space or container with its broadly drawn shape and hunky flowers in perfectly symmetrical cones. The deep green, waxy leaves are borne from reedy stems in a whirling spiral arrangement topped with very hot, very red cones that are actually bracts. The protective cones finally let loose their grip and burst open with true flowers that are pink and yellow. They protrude like imps sticking their tongues out in a laughable display. Dwarf cone ginger is a superior plant in garden beds and can be the easiest ginger to grow in containers year-round. The pots summer well in a mulched bed, or the gingers can be planted outdoors and lifted after they bloom. Wherever it is grown today, this Central American native brings swagger and bravado to the scene.

Plant Family
Costaceae

Other Common Name
Red button ginger

Bloom Period and Seasonal Color
Yellow-green to red bracts with small yellow and pink flowers in the summer and fall

Mature Height × Spread
Up to 5 ft. × 2 ft.

Dwarf Schefflera
Schefflera arboricola

Native to Taiwan, the so-called dwarf schefflera is not too small to make a big impression. The moniker refers to the size of its leaflets, which are quite small compared to other members of the clan. The plants are not from Hawaii, either, though their other common name of Hawaiian schefflera implies it. Dwarf schefflera can be the easiest tropical plant to grow, unless the gardener has a heavy hand with the water. Schefflera has dark evergreen leaves are oval and pert, densely arranged in round leaflets on green stems that turn warm gray-brown with age. Older plants use aerial roots like ropes to anchor the vine-y plant into its container or ground space. In the tropics and warmest subtropics, dwarf schefflera uses this habit to great effect as an espalier on a warm patio wall. The effect is a robust, shrubby green plant with muscular form that lights up its place in an indoor or outdoor garden.

Plant Family
Araliaceae

Other Common Name
Hawaiian schefflera

Bloom Period and Seasonal Color
Evergreen leaves

Mature Height × Spread
Up to 15 ft. × 10 ft.; averages 5 ft. × 5 ft.

When, Where, and How to Plant
Versatile schefflera thrives in low light and can be grown all year in a container indoors or happily spend summers in that container sunk into a deep mulch bed outdoors. They can be planted directly into garden beds for the warm months, dug up in the fall, and repotted for the winter. Avoid direct sun, indoors or outside. Provide an organic, well-drained soil by amending garden soil and potting mixes with compost or ground bark. Plant one in a standard 10-inch pot or space several 8 to 12 inches apart in rows and beds. Mulch schefflera plantings lightly, with no more than 1 inch of organic material such as ground bark or pine straw. If garden soils are likely to remain saturated, grow schefflera in pots.

Growing Tips
Allow dwarf schefflera to dry out between waterings. Overwatering can cause root rot and leaf drop. Fertilize monthly all year with a soluble formula while plants are growing. Once they reach the desired size, reduce fertilizer to four times annually. The plants will continue to grow as long as their roots are not constricted; repot small plants each year. To maintain a potted schefflera at a particular size, slip the plants out, root prune, and repot in the same container with fresh soil.

Care and Propagation
Be vigilant about mealybugs. Propagate by cuttings if you want to grow more plants.

Companion Planting and Design
Grow dwarf scheffleras with bamboo palms, false aralia, and jade plants for the entire green spectrum in one compatible display. Use it as a centerpiece with crotons, bush lilies, and night blooming cereus.

Try These
Variegated dwarf schefflera is painted with creamy yellow. *S. actinophylla*, umbrella tree, is a huge plant with big leaflets that has become invasive in Florida.

Earth Star
Crypthanthus hybrids

When, Where, and How to Plant
Earth star may be most successful outside the wet tropics as a container plant. Nearly any bright space indoors can host earth stars. Like their Bromeliad relatives, these little stars are surprisingly tough and are unaffected by dry air and drafts. Outdoors, find a site in dry shade or dappled sun where the plants can be watered but are protected from thunderstorms. Nestle them into the soil or sink their pots in a shallow mulch bed. Earth stars are not drought tolerant but neither can they stand saturated soils. Space plants 6 to 8 inches apart in groups or grow single earth stars in standard 6-inch clay pots. Take care when planting not to bury the center of the rosette. Do not mulch pots to maintain air circulation around earth star's very short stems.

Growing Tips
Overwatering can cause earth stars to look melted, flat, and limp. Water the plants and then let the soil dry out slightly between irrigations. Fertilize them regularly all year with a complete soluble fertilizer mixed at half strength every other month. If leaf tips look burned, move earth stars into a shadier location or use less fertilizer.

Care and Propagation
There are few pests to bug earth stars. Propagate by potting up offsets. See "Propagation" (page 36) for details.

Companion Planting and Design
A shallow planter can host a collection of earth stars or they can be the floor of a design including bush lilies, night blooming cereus, and snake plants. It is an excellent groundcover for dry shade.

Try These
Cultivars of note among the scores available are 'Chickadee', 'Pink Floyd', 'Houston', and 'Glad'.

Part of the allure of the tropical garden is its surprises and earth star, a native of Brazil, has more than its share. Tropical wonders might be an outrageous range of colors, patterns, and shapes that defy explanation; flowers that last for weeks; or plants hidden from view until their day comes. A few prized plants, including earth star, delivers all these qualities in one small package. Pink flowers are expected, yes, but earth star has dark pink leaves that look tender and wavy but are actually stiff and crinkled. Leaf patterns with expressionistic motifs and stripes down their length are certainly bold, but earth star has horizontal bands across its leaves. Tropical plants are noted for their huge, impressive statures, but the amazing earth star hugs the ground waiting for the gasp that follows its happy discovery. Terrestrial bromeliads, like the rest of the family, make excellent gift plants because they are virtually maintenance free.

Plant Family
Bromeliaceae

Other Common Name
Starfish plant

Bloom Period and Seasonal Color
Maroon, green, and cream striped leaves all year; small white flowers in the summer

Mature Height × Spread
Up to 8 in. rosette

Edible Pineapple
Ananas comosus

Be patient with pineapple plants! It may take up to two years to grow your own fabulous fruit, but in the meantime, this plant is unsurpassed as an icebreaker. No one can pass by without asking about the long, stiff, strappy, spiny leaves, and the variegated form is even more eye-catching. Once the flower spike does appear on a mature pineapple plant, the desire to see it fruit overtakes the gardener—and all piña colada fans in the neighborhood. They gather to see the flowers open grandly, first at night and then slowly over about twenty hours, then wait again for the fused cone of sweet fruit to mature. Pineapples have been grown and distributed throughout the tropical world by indigenous peoples for centuries. This noble fruit traveled from the New World home to Europe with Columbus in the 1500s as a fitting trophy of the bounty waiting across the ocean.

Plant Family
Bromeliaceae

Other Common Name
Piña

Bloom Period and Seasonal Color
Red or purple flowers inside green, yellow, or red bracts on mature plants, followed by fruit

Mature Height × Spread
Up to 3 ft. × 5 ft. in the garden and in containers

When, Where, and How to Plant
When a pineapple plant has eighty leaves, it is ready to bloom and fruit. Help it reach that goal with warmth and plenty of sunlight, six to eight hours in temperate zones. Prepare a well-drained, fertile garden soil or potting mix for young pineapple plants. Clay pots are preferred for their ability to allow water to evaporate through the surface. Grow each plant individually in a large clay pot, wide enough to prevent tipping. Indoors, a sunny warm room with only average humidity can host pineapple plants, but move the pots outdoors in summer. Protect plants by wrapping them in a towel or sheet during potting. Space plants 3 feet apart in a sandy greenhouse bed.

Growing Tips
Pineapple plants are not drought tolerant, but they should be allowed to dry out *slightly* between watering (so the surface is dry to the touch). Use warm water in cold weather, and do not allow water to accumulate in saucers underneath the pots. Fertilize monthly with a soluble complete formula to ensure steady growth and timely flower and fruit development. If plants with eighty leaves do not bloom, speed them along by putting the entire plant in a thin plastic bag with a cut apple to stimulate flowering.

Care and Propagation
Watch for mealybugs and scale insects. Propagate by fruit crowns, pups (or hapas) around the base of stems, and suckers between leaves.

Companion Planting and Design
Grow one plant in a 5-gallon container and use in a group of potted plants with similar needs, such as pink powder puffs, hibiscus, and devil's backbone.

Try These
'Smooth Cayenne' is green and produces three- to five-pound yellow fruits. Variegated pineapple has white-edged leaves and bright red fruit with white flesh. Gorgeous!

Elephant Ear
Xanthosoma sagittifolium

When, Where, and How to Plant

Elephant ears can thrive in shade, part shade, or dappled sun. Their growth potential is greater with increasing sun, as will be their need for water. Prepare a garden soil or potting mix that is richly organic and well drained. Use ground bark, compost, or compost/manure to improve existing conditions. Outside the tropics, start the tubers indoors several weeks before the last frost or purchase small plants. Space elephant ears 2 feet apart in beds or provide a container 10 inches wide and deep for each one. Mulch beds and pots with 2 inches of organic material such as ground bark. Plan to repot when roots creep out of the drain hole or when water rushes through the pot, indicating the roots have filled it.

Growing Tips

Water elephant ears very regularly using soaker hoses and timed irrigation for beds and pots. For fastest growth and largest leaves, do not let the soil dry out between irrigations. Fertilize when the first leaves appear and monthly after that until midsummer. When the mulch around elephant ears rots, work it into the soil and replace it. Allow the leaves die back on their own in the fall, and then dig up the tubers to store them, except in frost-free climates.

Care and Propagation

Elephant ear is another tropical with few pests to bug it. Propagate by dividing the roots and tubers.

Companion Planting and Design

Grow elephant ears with upright elephant ears and taro for a heart-shaped display of leaves and colors. Contrast them with African mask edged with nerve plant. Add them to water features for an instant tropical flair.

Try These

'Lime Zinger' has yellow-green leaves and 'Chartreuse Giant' shows its namesake color in stalks and leaves. Where this plant is outlawed, grow *Alocasia* species instead for a similar effect in the landscape.

Elephant ear is truly an international bon vivant. It is considered native to the American tropics, but has been carried so far and wide that its exact origins remain in dispute. Plants may need phyto-sanitary certificates of health when crossing borders, but elephant ear seldom needs an introduction anywhere. Indeed, it is categorized as invasive in central and south Florida and should not be planted there. Once the first leaf has opened on a tall, fat stalk, even non-gardeners can recognize its massive arrowheads, known botanically as a "sagittate" shape. Visually stunning, elephant ears are an unusual shade of deep blue-green with paler green on the lower leaf surfaces. This plant is entirely herbaceous, leaves with no trunk or woody stems. From one tuber comes a huge dome of foliage. One tuber creates an instant focal point in any water feature, but a row of elephant ears can line a pond bank or a driveway with equal ease.

Plant Family
Araceae

Other Common Name
Yautia

Bloom Period and Seasonal Color
Spectacular evergreen leaves

Mature Height × Spread
Up to 5 ft. × 5 ft.

Fan Flower
Scaevola aemula

Fan flower may sound like a demure plant, conjuring images of a shy smile hidden behind one in a lady's hand. Indeed, each flower is five petals arranged in a flat cluster that is also fan-shaped. The plants are low-growing and spread to form a shallow mound covered in flowers that are almost perpendicular to their stems. This radiating effect creates layers of flowers that look like delicate lace from a short distance away. Nothing could be further from the truth, since these plants are nearly bulletproof once they are established in garden beds or pots. In garden trials everywhere, recent introductions have received high marks for long-lasting flowers and durability. Thick stems and toothy leaves resist wilting in heat and rebound quickly from thunderstorms to make these natives of tropical Asia and Australia as welcome in the garden as they are beautiful.

Plant Family
Goodeniaceae

Other Common Name
Fairy fan flower

Bloom Period and Seasonal Color
Blue or purple and white flowers from the spring through fall

Mature Height × Spread
Up to 1 ft. × 3 ft.

When, Where, and How to Plant
Choose a site in full sun or mostly sun that has a reliable water source readily available. Prepare a soil for garden beds and containers that will be organic and fertile with very good drainage. Amend existing soils and potting mixes with sand and compost or other organic matter to achieve these conditions. Space young plants 12 inches apart in beds and mixed planters or plant two or three fan flower plants in a 10-inch pot that is wider than it is tall. These plants flourish in garden soil in warm weather, root easily, and are not difficult to maintain over the winter as potted plants in the temperate zone. Mulch plantings with 1 to 2 inches of organic material such as ground bark.

Growing Tips
Water and fertilize young fan flower plants regularly and let them dry out slightly between irrigations. Well-established plants will develop some drought tolerance. Use a general-purpose garden fertilizer as often as directed from the spring through fall. Maintain organic mulch around the plants and work it into the soil as it rots naturally. Deadheading flowers is not necessary, but tip pruning to remove a few sets of leaves can be done anytime the plants need refreshing.

Care and Propagation
Watch for slugs and snails on young plants, and spider mites in very dry conditions. Fan flowers can be propagated by cuttings.

Companion Planting and Design
In small spaces or mixed pots, combine fan flowers with scarlet sages, polka dot plants, and coleus. Let them spread around the base of larger plants like moonflowers and Mexican flame vines or African gardenias.

Try These
'New Wonder' is blue-purple, 'Bombay Blue' is dark blue, and 'Whirlwind White' is white.

When, Where, and How to Plant

Beds of ferns can line a shady walk year-round in the tropics and warm subtropics, where they will return from their roots even if the top is burned by cool temperatures. Farther north, they can be garden plants from the spring through fall, potted plants in the winter, or hang in a basket all the time. Choose a shady space with bright light but no direct sun for ferns. Indoors, a tabletop several feet away from a sunny window works well. Prepare a soil that is fertile, organic, and well drained by adding compost and ground bark to existing garden soils or potting mixes. Plant one fern in a 10-inch plastic basket or other similar sized container. Space fern clumps 1 foot apart in beds for thick stands.

Growing Tips

Ferns can dry out and drop fronds if the interior of their dense clumps is not well watered or if the air is very dry. Once monthly, fill a bucket with water and fertilizer and immerse potted ferns (pot and all) for several hours, more if the plant is very stressed. Use a soluble fertilizer with a balanced formula and occasionally add a layer of compost to the top of the soil around the fern. Repot annually, or when the stolons creep over the sides of their pot.

Care and Propagation

Watch for slugs and snails when growing ferns. If you want more, they can be propagated by division.

Companion Planting and Design

Ferns are the perfect complement to any shade-loving plant. Grow ferns with golden brush gingers and white bat flowers for contrasting shapes and colors.

Try These

Look for fine *Nephrolepsis* ferns like 'Fishtail', 'Kimberly', and 'Macho', but avoid 'Ladder' fern in the tropics.

More than twenty-five species form this genus of delightful terrestrial ferns that are native around the world in subtropical and tropical climates. Beginning with the ever-popular Boston fern, the Nephrolepsis *genus has found its way into virtually every garden, balcony, and dorm room. To further expand its domination of the fern world, both natural and man-made hybrids have increased the frond types, sizes, and colors that are widely available. These are plants with a deceptively delicate appearance that mask some very tough cookies. These ferns grow in clumps of pinnate, or cut, leaves that multiply by sending out stolons to creep and rapidly increase their size. The fronds are long, shaped like spears, and range from strong profiles like 'Macho' to fringed and frilly like 'Teddy Junior'. More than any other, these ferns are easy to grow and deliver reliable, bold, rounded shapes in a range of green hues.*

Plant Family
Nephrolepidaceae

Other Common Name
Sword fern

Bloom Period and Seasonal Color
Evergreen light to medium green leaves

Mature Height × Spread
Up to 3 ft. × 3 ft.

Firecracker Flower
Crossandra infundibuliformis

Abundant green leaves, shiny with deeply marked veins, cover the firecracker flower, a native of southern India and Sri Lanka. The leaves are arranged in a lateral form, reminiscent of dogwood but more densely layered, like most tropical plants. They jut out like strong chins determined to fulfill a mission—in this case, to show off magnificent flower clusters. The blooms are shaped like funnels with five lobes that spread out flat and almost upturned like a crooked smile. Bunches of the unusually angular flowers top each stem, held up above the leaves to be admired. Their colors are saturated yet subtle for the tropical palette in apricot, salmon, and pinkish shades as well as sunny yellow and scarlet red. Dynamic, charming, and unusual enough to be real conversation pieces, firecracker flower is small but powerful in raised beds, window boxes, and baskets.

Plant Family
Acanthaceae

Other Common Name
Crossandra

Bloom Period and Seasonal Color
Salmon, pink, yellow, or red flowers year-round

Mature Height × Spread
Up to 2 ft. × 2 ft.

When, Where, and How to Plant
Consider growing these flowering plants in containers year-round except in the dry tropics. Indoors, provide bright light for firecracker flowers and keep them away from drafts and heating vents. Outdoors, find a site in part shade or dappled sun. Grow them in raised beds or sink containers into beds of mulch in warm weather. Prepare a soil that drains very well and is richly organic by amending garden soils and potting mixes with organic matter. The plants are not drought tolerant but neither can they stand wet feet or flooded conditions. Growing in a ontainer offers the option of moving the pots indoors in rainy or cold seasons. Space firecracker flower plants 1 foot apart or provide a standard 10-inch pot for each plant. Mulch plantings with 1 inch of organic material such as ground bark.

Growing Tips
Overwatering can cause firecracker flowers to have straggly growth and fewer flowers. Allow the pots or plantings to dry out slightly between watering. Fertilize regularly with an all-purpose formula, preferably one with major and minor elements and trace minerals included. Mix a soluble product with water and apply monthly. Deadhead flowers unless you want them to produce seeds. As organic mulch decomposes, work it into the soil and replace it.

Care and Propagation
It has few pests, but watch out for spider mites. If you are into propagating, take cuttings and grow from seeds. See "Propagation" (page 36) for details.

Companion Planting and Design
Collect the set of three colors, or grow firecracker flowers with bird's nest ferns and peacock plants for upright interest. Plant it alongside impatiens and jacobinia for a contrast in flower shapes.

Try These
'Tropic Flame' has stunning coral flowers and one cousin has rare turquoise flowers.

Flamingo Flower
Anthurium spp.

When, Where, and How to Plant

Choose a warm, bright shady location outdoors to grow flamingo flower as an annual or to place pots for the summer. If the shade under a small, water-loving tree is dappled with a little sun, try it, but be prepared to water more often. In beds and in pots, prepare a soil that is richly organic and well drained. Amend native soils and potting mixes with organic matters such as compost or ground bark to enrich them and improve drainage. Use one-third as much organic matter as other planting material. Grow one plant in a 6- to 8-inch pot or space plants 10 inches apart in garden beds. Keep mulched with an organic mulch (other than pine straw), and work the mulch into the bed or pot as it decomposes.

Growing Tips

Flamingo flowers are moderate in their nature and react poorly to extremes. Water and fertilize very regularly in pots and beds. Use a complete balanced soluble fertilizer mixed at half strength once a month, but occasionally substitute a flower formula. If they are grown in garden beds outside the tropics, flamingo flowers must be potted up for the winter. Pots can grow year-round in a bright, humid indoor space away from direct sun but they benefit from time spent outdoors in warm weather.

Care and Propagation

Watch for slugs and snails on young plants. Propagate by dividing mature clumps or by starting seeds.

Companion Planting and Design

Flamingo flowers provide drama alone and shine when paired with Persian shield and peacock plants. Bird's nest ferns and chocolate plants offer boldly contrasting leaf shapes and colors.

Try These

Anthurium hookeri, bird's nest anthurium, forms more of a rosette. *A. crystallinum*, strap leaf, has huge leaves and white flowers with a yellow "tail."

The presence of flamingo flowers on the table naturally makes guests yearn for tall island drinks, if not a complete luau. A bouquet of bright red flamingo flowers might seem an offbeat choice, but in the hands of a St. Valentine's Day bride, these versatile, long-lived blooms are perfect. Shaped like hearts, their colorful spathes deservedly grab all the attention. They shine in riotous shades of red, pink, white, salmon, yellow, orange, and even green. The spathes are modified leaves that resemble wall sconces, meant to serve and shield the true flower. Those pop out along the tail, a rocket-shaped spadix that is usually white or cream. The mother of today's flamingo flowers is Anthurium andraeanum, named for the Greek words for "tail flower." Florists around the world cherish a rainbow of hybrids for their sheer tropical élan and tough good looks as cut flowers.

Plant Family
Araceae

Other Common Name
Patent leather plant

Bloom Period and Seasonal Color
Red, white, pink, and salmon flowers throughout the year

Mature Height × Spread
Up to 2 ft. × 2 ft.

Fragrant Heliotrope
Heliotropium arborescens

Tropical plants are defined by unusual leaves, stunning flowers, fragrance, and strong shapes in the landscape—but not all of them possess these qualities in equal measure. Fragrant heliotrope leads the pack on tropical assets, a plant deserving of its reputation as unforgettable and evocative. The leaves are thick, oval, and almost ridged with a strong pattern of sunken veins. The veins run tightly across the leaf from rib to edge and cause puckering, like embroidery thread drawn too tight for its fabric. All those crevices trap every drop of rainfall available to assist the impossibly sweet-smelling blooms. Each flower head holds a cluster of stems crowned by dozens of regal purple, trumpet-shaped flowers. The stiff stems hold the flowers proudly upright above the leaves, all the better to send their fruity fragrance into the air of their native Peru and into gardens large and small.

Plant Family
Boraginaceae

Other Common Name
Cherry pie plant

Bloom Period and Seasonal Color
Fragrant purple flowers in the spring and summer

Mature Height × Spread
Up to 3 ft. × 3 ft.

When, Where, and How to Plant
Fragrant heliotrope thrives in fertile, organic, well-drained soil with a reliable irrigation source nearby. Provide suitable soil for beds or containers by amending existing garden soils and potting mixes as needed with organic matters like compost and aged manures. The soil for heliotropes should be able to hold some moisture and tolerate watering regularly without developing a saturated root zone. Heliotropes thrive in morning sun with some shade in the afternoon, such as that provided by taller plants. Space young plants 10 to 12 inches apart in beds or large mixed planters, or plant two in a standard 10-inch pot. Water new plantings well and mulch beds and pots well with 1 to 2 inches of organic material such as ground bark.

Growing Tips
Heliotrope plants will be stunted and fail to bloom without a regular schedule of deep watering. Pinch young plants two or three times to encourage branching and more flower heads. Fertilize heliotrope regularly. Use a complete formula at the recommended rate from the spring through summer and at half strength in the winter. Maintain a mulch blanket around the plants and work rotting mulch into the soil. Deadhead fading flowers to stimulate more blooms.

Care and Propagation
Watch for snails and slugs on young plants, and whiteflies in summer. Propagate by cuttings and seed, if you want to grow more. See "Propagation" (page 36) for details.

Companion Planting and Design
Grow heliotropes with reed stem orchids, red gingers, and passion flowers for deep jewel tones of red and purple. Let them bring some contrast to Philippine orchids, sanchezias, and papaya plants.

Try These
'Marine' heliotrope is compact and stocky with long-lasting fragrance.

Gazania
Gazania rigens

When, Where, and How to Plant

Pick a sunny site for this stalwart plant, which can grow outdoors year-round in the warmest subtropics as well as in its native tropics. Gazanias grow best in soil that is moderately fertile yet well drained. The plants are not entirely drought tolerant but cannot tolerate flooding either. Prepare a soil in beds or containers that combines garden soil or potting mix with sand and organic matters to improve their fertility and drainage. Transplant small gazanias in the spring or summer and be certain they are resting at the same level they were growing originally. Space plants 8 inches apart in beds or group three gazanias in a standard 8-inch pot. Water new plants regularly until they become established. Mulch lightly with no more than 1 inch of organic material.

Growing Tips

Provide a regular program that allows gazania plants to dry out between waterings. Use a complete, all-purpose fertilizer regularly from the spring through fall and in minimal amounts in the winter. Deadhead the flowers to stimulate the development of new buds. Outside the tropics, lift plants in beds and pot up to spend the winter indoors. Always inspect the dense foliage for insects before making that transition. Keep mulch to a minimum in beds.

Care and Propagation

Watch for slugs and snails on young plants, and spider mites in very dry conditions. If you want to propagate, do so by seeds.

Companion Planting and Design

Gazania belongs with equally bright flowers like hibiscus, mallows, and dwarf powder puff plants. Grow it as living mulch with pineapple plants or hoya, or let it fill the space around a miniature date palm.

Try These

'Tiger Stripe Mix' has orange-striped flowers, 'Creamsicle' is tan, and 'Daybreak Pinks' is pink.

Native to South Africa, gazania dances across the hottest garden spots with cool green or silver leaves and eye-popping flowers. Its native origins demand much of a plant, and gazania delivers with drought and heat tolerance plus colors bright enough to attract pollinators blinded by white hot sun. Their leaves are skinny and finely cut and can look like ankle-high grass from across the garden. They shimmer in the sun, a drum roll for the approaching host of teacup-sized flowers. The color spectrum of gazania flowers is most unusual, ranging from rich cream and rosy pink to stunning yellow, gold, and orange. The solid tones are dazzling jewels with prominent centers, while the bicolor flowers are raging sunbursts. Their hot pops of color and texture are equally at home in garden beds as edging or groundcover and in containers alone or with mixed company. A note of caution: Gazania has become a pest plant in California.

Plant Family
Asteraceae/Compositae

Other Common Name
Treasure flower

Bloom Period and Seasonal Color
Orange, yellow, and cream flowers primarily in the summer and fall, sporadically in the winter and spring

Mature Height × Spread
Up to 1 ft. × 1 ft.

Gerber Daisy
Gerbera hybrids

There's something to be said for relentless good nature, and gerber daisies keep their sunny side up. Even when the plants are not in bloom, their crinkled leaves turn up their edges in a cheerful smiling clump. Bright green stems hold the flowers aloft, a rainbow of fringed daisies with deep yellow centers. In a further display of cockeyed optimism, the buds curl slightly to one side so the flowers do too. Gerbera jamesonii and G. viridifolia are the parents of the myriad gerber daisy varieties and selections available everywhere as cut flowers and garden plants. As in their native South Africa, warm sun, good drainage, and a ready water supply are the keys to success with this plant and its progeny. Wherever those conditions exist or are created, the plants thrive and bloom on endlessly. Clumps of gerber daisies spread but not rapidly and can become reliable perennials in the subtropics and a bit farther north in well-drained garden soils.

Plant Family
Asteraceae/Compositae

Other Common Name
Transvaal daisy

Bloom Period and Seasonal Color
Flowers appear all year in every color except blues

Mature Height × Spread
Up to 1 ft. × 1 ft.

When, Where, and How to Plant
Choose a site in sun or part sun for gerber daisy, which can grow outdoors year-round in the warmest subtropics as well as in its native tropics. The plants can be lifted in the fall and potted up to overwinter indoors elsewhere. The soil should be well drained, fertile, and organic. Gerber daisies are not drought tolerant but they cannot tolerate wet feet either. Prepare a soil for beds or containers by amending garden soil or potting mix with sand, compost, and other organic matters. Space plants 8 inches apart in beds or group three gerbers in a 12-inch pot. Water new plantings well and regularly until established. Mulch plantings with 1 to 2 inches of organic material such as ground bark.

Growing Tips
Put gerber daisies on a regular regime that provides plenty of water for strong leaves and flowers but allows the plants to dry out between irrigations. Put a granular, complete formula, all-purpose fertilizer around the base of the plants four times each year or use a soluble formula every other month. Keep mulch to a minimum in beds and repot container plants annually as needed to accommodate growing roots. Cut flowers or deadhead as they fade.

Care and Propagation
Few pests bother gerber daisies. Propagate them by division and cuttings.

Companion Planting and Design
Grow gerber daisies with China asters for cut flowers. Design an orange-red garden with gerbers, gazanias, and kanga paw. Plant a row in front of bottlebrush and edge it with Cape daisies.

Try These
Dozens of hybrid gerber daisies are widely available and grow as easily as their parents.

Hibiscus

Hibiscus rosa-sinensis

When, Where, and How to Plant
Choose a sunny site for this outstanding tropical plant, outdoors in beds or pots and indoors in sunrooms. Avoid sites where wind and salt spray predominate, and do not crowd plants against solid walls. Hibiscus grows best in soil that is organic and fertile yet well drained. Leaves will become pale and flowers will be small without adequate amounts of both water and sunlight. Prepare a soil for beds and pots that combines garden soil or potting mix with finely ground bark or sand, compost, and other organic matters. Plant hibiscus 2 feet apart in beds or large planters or in containers at least 10 inches across and deep. Plant at the same level or slightly higher than the hibiscus was growing originally.

Growing Tips
Maintain hibiscus on a routine that allows the plants to dry out slightly between waterings. Fertilize regularly with a granular or soluble formula year-round. Clip back stem tips to keep its growth bushy and to prepare for moving it indoors for the winter. Keep mulch to a minimum in pots and beds. Repot annually as needed to provide ample root space and to prevent tree forms from tipping over. If the leaves turn pale, move hibiscus to a location with more sunlight.

Care and Propagation
Be on the watch for aphids and whiteflies. Propagate by cuttings if you'd like more hibiscus.

Companion Planting and Design
Grow hibiscus with equally large plants with diverse textures like dwarf powder puffs and bottlebrush. Surround it with China asters, kanga paw, gazanias, and Cape daisies for complementary and contrasting flowers.

Try These
Cajun hibiscus has glossy leaves, huge flowers, and can become perennial in the subtropics.

Within many pieces of music, there is a phrase that repeats itself, calling out for a response from another instrument within the same piece. When the response raises the bar of technique and musicality, the piece becomes a "groove." Hibiscus called out long ago from its native China, and the response continues in the most colorfully diverse group of flowers to be found anywhere. Easily recognized, the classic lobed bracts look starched, open wide to expose the prominent tongue like a broad smile on a peevish child. Every hue of red, pink, orange, and yellow can be found in endless combinations of single, semi-double, and double flowers on modern hibiscus. Solid color flowers, bicolors, ruffles, and painted flourishes have been developed from this simple plant and its relative *H. schizopetalus*, a native of East Africa. Hibiscus called, and humans are still responding with new variations released every year.

Plant Family
Malvaceae

Other Common Name
Chinese hibiscus

Bloom Period and Seasonal Color
Red, orange, yellow, or pink flowers year-round

Mature Height × Spread
Up to 15 ft. × 5 ft. in tropics; averages 6 ft. × 3 ft.

Hidden Ginger

Curcuma longa

Hidden ginger holds a special place in the tropical garden because it is different from its kin in the ginger family and from other tropical plants too. While the others are show-offs with plenty to display, this one hides its light under a bushel, or in this case, long corrugated-looking leaves. Each leaf is a sword that springs from an underground rhizome that is ground up for use as a spice (turmeric) and a stunning yellow dye. The flowers are secreted deep in the leaves and seldom can be seen at all without lifting the foliage or lying down next to the plant. They are worth the search—sweet little stalks packed thick with blooms that look for all the world like pine cones with neon lights in them. The shape is right but the waxy flowers are reddish pink at the bottom and yellow at the top.

Plant Family
Zingiberaceae

Other Common Name
Turmeric

Bloom Period and Seasonal Color
Bracts fade from green at the bottom to white on top with a rose tinge in late summer to fall

Mature Height × Spread
Up to 3 ft. × 3 ft., usually smaller

When, Where, and How to Plant

Hidden ginger thrives in part shade or dappled sunlight with plenty of water whether it is grown in pots or garden beds. The plants need richly organic soil that can hold water almost constantly. Prepare a soil with these qualities by blending garden soil or potting mixes with organic matters such as compost and ground bark in nearly equal amounts. Plant clumps 1 foot apart or give a solo plant its own 12- to 14-inch container. Keep beds and pots mulched lightly with 1 inch of organic matter such as ground bark. Work decomposing mulch into the planting area regularly to replenish its nutrition and organic content.

Growing Tips

Dry leaf tips on hidden gingers can result from too much sun or too little water. Water very regularly and do not allow the soil to dry out, but do allow it to drain well between irrigations. Consider soaker hoses to simplify this task if needed. Fertilize plants in the spring and summer with an all-purpose formula that includes both major and minor elements. Divide this once-a-year bloomer right after it flowers. Replant the clumps immediately since they die down naturally in late fall and can be difficult to locate until they emerge in the spring.

Care and Propagation

Few pests ever become a problem. Increase your collection of hidden gingers by dividing its clumps.

Companion Planting and Design

Grow hidden gingers among ferns and split leaf philodendrons with a skirt of nerve plants. Pair it with prayer plants for contrast in form and color, or with African mask for one very exotic pot plant.

Try These

'Panama Purple' has deep purple flowers in the spring and dark stripes on the leaves.

When, Where, and How to Plant

Take into consideration that *all parts of this plant are toxic* when selecting a site. Select sunny locations near a reliable source of water necessary for steady growth and luscious flowers. These plants like water but cannot tolerate wet feet. Prepare a soil for beds and containers that is well drained, organic, and fertile. Where native soils are heavy, dense, or very sandy, amend with a combination of organic matters. Space plants 3 feet apart in beds. Test drainage in potting mixes before planting and amend if needed. If you're growing in pots, provide a container 12 inches wide and deep. Transplant at the same level the plant was growing originally and soak deeply. Mulch with 2 inches of organic material such as ground bark. Stake young plants if needed but remove the supports once a plant can stand on its own.

Growing Tips

Leaves as large as these need water regularly, with only a brief opportunity to dry out between irrigations. Soaker hoses, ground level sprinklers, or drip systems on timers can make this task more routine. Fertilize regularly with an all-purpose garden formula that has major and minor elements and trace minerals in it. Maintain 2 inches of mulch around the plants in garden beds and pots and work it into the soil as it decomposes. Clip off seedpods in warm climates to avoid rampant reseeding.

Care and Propagation

Watch out for slugs and snails on young plants. Propagate by tip cuttings in late summer.

Companion Planting and Design

Blanket the soil around horn of plenty with licorice plants. Go for strong shapes and texture in companions like chenille plants, taro, Chinese violets, and Brazilian red cloaks.

Try These

Angel's trumpet (*Brugmansia*) resembles this relative, but its flowers hang straight down.

If the Mexican native horn of plenty plant were not so clearly a three-dimensional plant armed with noisy ruffles and flourishes, it could be a cartoon character. It might be an imaginary plant, made up for a kids' movie, since nowhere else would a living thing be drawn with such incongruity. Horn of plenty leaves are an uptown version of its kin in the tomato-eggplant family, but its fruits are thick-walled, full of fuzz and seeds, and covered in green bristles. The plants grow rambunctiously but do not climb and can be rather a messy, if charming, mound. The silky flowers are exquisite, long-tubed trumpets topped by fat corkscrew buds. Twisted into pinwheels with precious recurved points, they sweep into perfect trumpets but retain the barbs at each lobe. The effect of so many huge white flowers and buds is as dazzling at high noon as it is under a full moon.

Plant Family
Solanaceae

Other Common Name
Devil's trumpet

Bloom Period and Seasonal Color
Purple and white flowers in the summer and fall

Mature Height × Spread
Up to 6 ft. × 4 ft.

Impatiens
Impatiens walleriana

Sometimes wonderful plants suffer from overexposure, like a lovely starlet cast one too many times in a bikini to be taken seriously in any other role. Such is the lot of impatiens, or sultanas, to be ubiquitous, planted everywhere shade limits other choices. Of course, these easy-to-grow plants personify the tropical attitude, which is why they are so popular. Medium or dark green leaves are heart-shaped and form dense clumps to show off the blooms. They are the fancy chorus line in the garden, a sea of blooming jewels done up in a rainbow of colors. Native to East Africa and cultivated since the last century, today's impatiens shine in saturated solids hues and bicolors, with single and double flowers. Darker flower tones often have darker leaves and can tolerate more sun, but the newest release from this family, 'Sunpatiens', grows in full sun.

Plant Family
Balsaminaceae

Other Common Name
Sultana

Bloom Period and Seasonal Color
Red, orange, purple, or white flowers in the summer

Mature Height × Spread
Up to 2 ft. × 2 ft.

When, Where, and How to Plant
Choose a location in warm, bright shade outdoors to grow this plant as an annual or to place pots for the summer. In beds and in pots, prepare a soil that is richly organic and well drained. Amend native soils and potting mixes with organic matters such as compost or ground bark to achieve these conditions. Use one-third as much organic matter as other planting material. Grow one impatiens plant in a 6- to 8-inch pot or space plants 10 inches apart in garden beds. Keep them mulched with organic material and work the mulch into the bed or pot as it decomposes. Young impatiens plants grow well indoors in a bright, warm room grouped with other plants to increase the humidity around all of them.

Growing Tips
Impatiens need a routine that supplies water and fertilizer regularly to garden beds and pots. Water the plants, then let the soil dry out only slightly before watering again. Fertilize monthly with a soluble formula and use a slow-release fertilizer in the summer. If they are grown in garden beds outside the tropics, the plants must be potted up or new cuttings rooted to grow indoors over the winter. Cut several inches off of plants anytime they become leggy.

Care and Propagation
Watch for slugs and snails on young plants, and spider mites in very dry weather. Impatiens can be propagated by cuttings and seeds.

Companion Planting and Design
Masses of impatiens in mixed or solid flower colors can fill beds or pots in shady places. Add a row to a bed of nerve plants and caladiums.

Try These
Double-flowered impatiens look like tiny roses and New Guinea impatiens (*I. hawkeri*) offer wildly striped leaves.

Jacobinia
Justicia spp.

When, Where, and How to Plant

Pots of jacobinia can grow year-round in a bright indoor space away from direct sun but benefit from time outdoors in warm weather. It is a fine plant for garden beds year-round in the tropics and warm subtropics and can become a returning perennial in sheltered locations slightly farther north. Choose a warm location in bright shade for beds or pots of jacobinias. Amend native soils and potting mixes with organic matters such as compost or ground bark to enrich them and improve drainage. Use one-third as much organic matter as other planting material. Space plants 1 foot apart in rows or beds or grow two jacobinias in a standard 10-inch pot. Mulch plantings with 2 inches of organic material.

Growing Tips

These plants react poorly to stressful extremes of drought and flood. Water and fertilize very regularly in pots and beds. Use a complete balanced soluble fertilizer mixed at half strength once monthly all year. If grown in garden beds outside the tropics, the plants must be potted up for winter. Deadhead flowers as they fade to promote more buds and cut down leggy stems. Work rotting mulch into the soil and replace it promptly.

Care and Propagation

Few pests ever bother jacobinias. Propagate by cuttings (with long stems and most leaves removed). See "Propagation" (page 36) for details.

Companion Planting and Design

Jacobinias are the star of any shade garden along with bird's nest ferns, peacock plants, and impatiens. Pair some with Persian shield or chocolate plants for stunning contrast.

Try These

Coral shrimp plant (*J. brandegeana*) is hardier but less showy, with coral bracts and white blooms.

Native to South America, jacobinia marches into gardens everywhere with strong lines of straight-up stems, sharply drawn leaves, and crisp, erect flowers. The plants bring a military swagger to shady spaces whether they are grown in beds or containers. Their leaves are longer than they are wide, coarsely veined, and stick out from their thick stems in a bright salute. Sizes range from knee-high mounds to shrubby specimens up to 6 feet tall. Most of the jacobinia plants available outside the tropics are hybrids with neat habits on middle-sized plants with flowers in a wide range of shades. The flowers seem papery at first glance, plumes created by dozens of long, recurving petals so perfectly crafted that they look artificial. Jacobinia flowers are the signal corps of the shade garden, waving their flags to attract insects eager for their dizzying nectar. Even after a thunderstorm, they stand at attention and bring comforting order to the shade garden design.

Plant Family
Acanthaceae

Other Common Name
Brazilian plume, justicia

Bloom Period and Seasonal Color
Pink, white, red, purple, orange, or yellow flowers primarily in the summer

Mature Height × Spread
Up to 6 ft. × 2 ft.

Jade Plant
Crassula ovata

Jade plant is certainly an oddball among the tropicals, a succulent tree form plant that thrives on modest inattention. The stems are gray, smooth, and husky looking, crowned by dark green, spade-shaped leaves that are thick with watery tissue. The effect is clean, geometric, and just strange enough to leave some people cold. Others are drawn to its unusual good looks and collect a family of solid and variegated leaves. Jade plant has spread all over its native South Africa because one thick, life-giving leaf can become a plant and its airy, starlike flowers produce viable seed. In similar conditions, jade plant is a large, chunky presence worldwide. Outside the tropics, it can be spectacular and ages gracefully into a venerable container plant with a bonsai feel. Jade plant is a good companion in dry, low-light offices where, although it is unlikely to bloom, it tolerates even dusty conditions with complete aplomb.

Plant Family
Crassulaceae

Other Common Name
Jade tree

Bloom Period and Seasonal Color
Evergreen, stubby leaves and pink flowers on mature plants

Mature Height × Spread
Up to 10 ft. × 4 ft. in native environs; most often less than 3 ft. × 3 ft.

When, Where, and How to Plant
Jade plant thrives in bright light or part sun outdoors in a dry tropical climate. Everywhere else, they are unbeatable in containers indoors all the time or outdoors in warm, dry weather. In rainy climates, put pots on a porch or under an eave to keep them dry. Prepare a soil that is fertile and very well drained by incorporating sand and ground bark into garden soils and potting mixes. Space small jade plants 6 inches apart or select a small container wider than it is tall for stability. Mix solid green-leaved types with variegated jade plants for added interest. Plant or pot a jade plant's stem at the same level it was growing originally; do not bury the stem. Do not mulch jade plants in pots or beds and do *not* put saucers under pots.

Growing Tips
Give jade plant a routine that varies little during the year. Water the plant, let the soil dry out, then water again. Poke your finger into soil and water when it feels dry up to your first knuckle. Use a soluble formula fertilizer mixed into the water every other month or use a comparable granular product three times a year. Repot in the spring as needed to provide space for the growing plant. Leaf drop reflects overwatering; shriveled leaves need water.

Care and Propagation
There are few pests, but watch for mealybugs on jade plants. Propagate by tip and leaf cuttings.

Companion Planting and Design
Jade plants deserve companions that are equally bold such as edible pineapples, foxtail agaves, and spineless yuccas. You can soften its impact by pairing it with musical note plants, gazanias, and dwarf powder puffs.

Try These
Dazzling 'Sunset' has yellow leaves with red margins. 'Tricolor' is green, pink, and white.

Kangaroo Paw
Anigozanthos flavidus

When, Where, and How to Plant
Plant seedlings or small plants in warm soil that is fertile, organic, and very well drained. Amend garden soils and potting mixes with sand or ground bark to make sure it has these qualities. Choose a sunny or partly sunny site with a south- or west-facing exposure for these heat lovers. Kanga paw is not drought tolerant; make certain the site has ready access to water. Space plants 8 inches to 1 foot apart in groups, rather than in rows, for the best effect. A group of three in an 8-inch pot makes a good show alone, but kanga paw is a good container companion. Water well after planting and apply a blanket of organic mulch around the plants. The plants may return from their roots in subtropical regions.

Growing Tips
Water kangaroo paw plants often enough to keep the well-drained soil well watered and to approximate its natural habitat. Use soaker hoses or well-timed irrigations and keep the bed mulched. Fertilize regularly with a complete flower formula. Use a granular product or a soluble fertilizer that is mixed in water as directed on the product label. Work the mulch into the bed or pot as it decomposes. Deadhead flowers as they fade to stimulate new buds.

Care and Propagation
Few pests hang around kangaroo paw. Propagate it by cuttings taken in the summer and from seeds planted in late winter.

Companion Planting and Design
Mass kanga paws so their flowers nod at the base of a bird feeder or use them with skyflowers, candelabra plants, and Chinese hat for fine contrasts in colors and flower shapes.

Try These
Look for these and many more: 'Pink Joey', 'Gold Velvet', 'Bush Diamond', and 'Big Red'.

At first glance, kangaroo paw is obviously a different sort of tropical plant. Its stems are interesting to look at—narrow, green, and a bit nubby. They rise from a strappy rosette of thin leaves, topped by bloom stalks of tightly curled, furry flowers arranged on each stem like a chandelier earring. Each flower is a tube with a puffy end, aligned with its mates on the stem in the shape of a kangaroo's paw. The fact that each tube has short, soft fuzz all over it adds to the plant's charm as well as its Down Under image. Kangaroo paw comes from Australia, where it grows in eucalyptus forests and on rainy riverbanks. It looks a bit rugged, with its geometric arrangement of fuzzy, knobby flowers. That quality and its lasting qualities make it a sought-after cut flower. Its overall effect is joyous and uplifting to the spirits.

Plant Family
Haemodoraceae

Other Common Name
Kanga paw

Bloom Period and Seasonal Color
Orange, yellow, pink, red, or green flowers in the summer

Mature Height × Spread
Up to 3 ft. × 1 ft. in the garden and in pots

Licorice Plant
Helichrysum petiolare

Every garden palette needs filler plants—those that quietly occupy a space to set off other, more striking plants in the design. Licorice plant would be such a plant wherever it came from, but the African native plays a role beyond its beautiful leaves and creeping form. Grown for its masses of leaves, licorice plant is supremely amusing all on its own, vigorous and fast-growing, though honestly overshadowed at times by its bedmates. The species has silver-green leaves with a sweet downy sheen on their surface and small yellow flowers. Selections range in leaf color from nearly yellow to quite gray on plants that mound up and out at the same time. All serve as living mulch to the plants around them, holding water in the soil and shading the roots to keep them cool. A more colorful and reliable groundcover could not be invented for the tropical garden.

Plant Family
Asteraceae/Compositae

Other Common Name
Hottentot tea

Bloom Period and Seasonal Color
Gray-green leaves in shades from dark green to chartreuse

Mature Height × Spread
Up to 12 in. × 24 in.

When, Where, and How to Plant
Licorice plant grows well in sun, part sun, or dappled sun as long as there is a reliable water source nearby and ready for use. Sunnier sites will require more water, but growth will be faster. Provide a well-drained soil that is organic and fertile for beds and containers. Amend existing soils with sand and organic matters such as compost and ground bark to achieve those conditions. Space licorice plants 12 to 14 inches apart in beds or large mixed pots or grow one in a standard 6- to 8-inch pot. They thrive in humid, warm weather outdoors in garden beds and can go through the winter as potted plants indoors in a bright, warm room. Water new plants well and add a layer of organic material such as ground bark.

Growing Tips
Water regularly and allow them to dry out only slightly between watering. Licorice plants are not drought tolerant, nor do they tolerate saturated root zones. Pinch young plants to encourage lots of growing tips to increase their spread. Use a general-purpose fertilizer regularly, such as monthly applications of a soluble formula. As mulch decomposes, work it in to the soil and replace with fresh organic material. Remove flowers to keep leaves growing.

Care and Propagation
Watch for chewing caterpillars and rust attacking licorice plants. Propagate by cuttings if you want more.

Companion Planting and Design
Licorice plants contrast well with scarlet sages, fan flowers, and cat's whiskers or use it as a filler plant in mixed pots. Grow it as living mulch around caricature plants, spineless yuccas, and a joy perfume tree.

Try These
Look for 'White Licorice', 'Lemon Licorice', and 'Licorice Splash', which has strong cream and chartreuse colors.

Lion's Ear
Leonotis leonurus

When, Where, and How to Plant

Choose a sunny site that is warm and away from excessive winds. In sites with fewer than six hours of sun, the stems will be weak and the flowers less abundant. Prepare a well-drained, fertile soil by amending existing garden soils and potting mixes with ground bark and sand, if needed. Space plants 1 foot apart in beds or mixed plantings, or provide a standard 10-inch clay pot for each lion's ear plant. Like other mints, lion's ear can be pinched back at planting to stimulate its branching and more flower stems. Mulch new plantings lightly, with no more than 1 inch of organic material such as shredded bark. Water young plants well until they are established, but take care not to overwater mature lion's ear plants.

Growing Tips

Lion's ear plants need water regularly until they are established in beds or pots, and then must dry out between irrigations. Use a soluble general-purpose fertilizer mixed in the water monthly, or a granular product as often as directed on its label. Prune mature plants if they become leggy or to stimulate new growth. Potted plants can be cut back as needed to overwinter them, or plants can be propagated and grown indoors. Deadhead flowers as they fade to make room for more.

Care and Propagation

There are few pests that bother these plants. Propagate by softwood cuttings and seed.

Companion Planting and Design

Make lion's ears an integral part of your butterfly garden by pairing them with bottlebrush and butterfly bush. Grow it with ti plant and devil's backbone for a textural riot, or surround it with Cape daisies and kanga paws.

Try These

'Harrismith White' is a cultivar with white flowers.

Native to South Africa, lion's ear carries the banner of its mint family heritage so vividly it could be a caricature of the group. The mint group, or Laminaceae, is noted for holding their leaves opposite each other on square stems. Some, like coleus, have so many leaves that their arrangement is blurred, but lion's ear displays stark stems with strappy leaves in opposite pairs at 2-inch intervals. Both stems and leaves are shiny porch green and rise dynamic and erect to face sunny, dry microclimates. In the leaf axils, the flower clusters start as exotic, ping-pong-ball-sized structures that look like sweet gumballs painted green with orange dots between their spikes. Those dots soon fling themselves into the air with gusto, hairy closed tubes about 2 inches long that someone thought resembled a lion's ear. The blossoms surround the stem in a whirling dervish of bizarre, fiery orange.

Plant Family
Lamiaceae

Other Common Name
Wild dagga

Bloom Period and Seasonal Color
Orange flowers in the summer and fall

Mature Height × Spread
Up to 6 ft. × 3 ft.

Lobster Claw
Heliconia stricta

Lots of tropical flowers are appropriately described as huge clusters of colorful petals with sumptuous fragrances, but only a select few can stand up to the crowd armed only with basic lines and primary colors. Lobster claw and other Heliconia species, along with bird of paradise plants, bring clean, upright leaves and flower stalks to the scene in their native South America and in gardens worldwide. These plants offer stunning contrast to the rest of the pack by providing strongly outlined shapes with well-known silhouettes. A branch of lobster claw in bloom looks like an exotic version of old-fashioned television antennas, thick straight rods with appendages lined up opposite each other all along their length. The true flowers present inside shell-shaped bracts are simple enough to be drawn in crayon, yet are as widely recognized as any company logo. Even one stem in a flower arrangement sets an unmistakable tropical tone.

Plant Family
Heliconiaceae

Other Common Name
Parrot beak

Bloom Period and Seasonal Color
Red or yellow flowers

Mature Height × Spread
Up to 8 ft. × 3 ft.

When, Where, and How to Plant
Lobster claw thrives in fertile, organic, well-drained soil with regular irrigation. The plants will be damaged when temperatures near 50 degrees F and so are often best grown in containers for the long term in areas that reach those temperatures. Prepare the soil in beds or containers by amending as needed with compost, ground barks, and other aged products such as manure that will maintain an acidic soil. Ideally, the soil must be able to hold its moisture and tolerate regular watering without causing the plants to stand in water. Flowers will form sooner in sunnier sites, but lobster claw plants adapt well to partly sunny areas and bright indoor areas. Be sure that young plants are at the same depth they were growing originally; do not bury the crown. Space plants 2 feet apart in beds or equally large planters. Group lobster claws with other water-lovers to simplify their maintenance.

Growing Tips
Water pots *very* regularly and use soaker hoses (or similar) to irrigate outdoor plantings. Keep plantings mulched to a depth of 2 inches in beds, 1 inch in pots. Fertilize lobster claws with a complete granular formula three or four times each year or use soluble formulas monthly from spring to fall (less often in winter). Work rotting mulch into the soil. Cut flowering stalks for the vase or when the flowers fade on the plant.

Care and Propagation
Watch for slugs and snails on young plants. If you want more of this spectacular plant, propagate by dividing the rhizomes.

Companion Planting and Design
Grow lobster claws with equally strong forms like ruffle palms and flaming glorybowers. Surround them with color and textural complements such as red gingers, reed stem orchids, and fragrant heliotrope.

Try These
The dwarf variety 'Jamaican' brings smaller sized red "claws;" 'Golden Torch' has yellow bracts.

Lollipop Flower
Pachystachys lutea

When, Where, and How to Plant

Pick a location in sun or mostly sun with a dependable water source nearby to grow lollipop flower in garden beds and pots. Prepare the soil by amending existing garden soil or potting mix with organic matters so it will be fertile with very good drainage. Space plants 12 inches apart in garden beds and mixed planters, or plant two small plants in a standard 12-inch pot. Be sure the plants will be growing at the same level they were originally; do not bury the stem. These plants flourish in garden soil in warm weather and are not difficult to maintain over winter as potted plants in the temperate zone. Take cuttings in summer or dig up plants in the fall. Mulch beds and pots with 1 to 2 inches of organic material.

Growing Tips

Water lollipop flower plants very well and then allow the soil to dry slightly between waterings. Use a general-purpose garden fertilizer regularly throughout the year to keep new leaves and flowers coming. You will know when to water because the soil will look dry on the surface when it needs more. Deadhead old flowers as they age by clipping the entire bract and the small, modified leaves below it. Maintain organic mulch around the plants and work it into the soil as it rots naturally.

Care and Propagation

Watch out for mealybugs. You can propagate more plants by taking cuttings. See "Propagation" (page 36) for details.

Companion Planting and Design

Perfect companions include jacobinias, bat face cupheas, polka dot plants, cat's whiskers, and yellow bells. Use these as an accent in front of vines like hyacinth bean and Chinese violet.

Try These

A plant with a similar waffled flower is the coral shrimp plant, *Justicia brandegeana*.

Forget the birthday candles? It's no problem if the rich yellow steeples of lollipop flower are in bloom, which they are likely to be. Native to sunny, wet microclimates in the Peruvian foothills, lollipop flowers are good soldiers indoors everywhere. They bloom almost year-round, waffle cones of bracts, bright candles lit up with white flames when the true flowers emerge from their depths. They stand up straight—like lollipops! in a candy store window, mounted on stiff stems above candelabras of dark green leaves. The foliage is deeply veined and looks rather like cardboard imprinted with a sober design—all the better to host the outlandish flowers. These startling, surprisingly rugged features are the source of its botanic name, which is Greek for "thick spike." The flower color of lollipops is unsurpassed even in the tropical palette—a saturated and cheerful yellow that lights up any room or garden space.

Plant Family
Acanthaceae

Other Common Name
Yellow shrimp plant

Bloom Period and Seasonal Color
Yellow flowers in the summer and fall

Mature Height × Spread
Up to 2 ft. × 2 ft.

Malaysian Orchid
Medinilla myriantha

Malaysian orchid is not actually an orchid at all, and it is said to be native to the Philippine Islands and perhaps Malaysia. Its enigmatic beauty belongs in the bright shade of a tropical garden that enjoys warm days and nights, as well as ample humidity. The plants create a rounded shape of rough good looks wherever they are grown. The leaves are big and bold, leathery, and gorgeous deep evergreen. They are deeply marked by graceful, curved light green veins that look stitched— quilted even—by Nature's own needles. Huge flower clusters look like bunches of rose-colored grapes at first, then burst open to reveal succulent, waxy bracts around yellow true flowers. The blooms appear over a period of months and each lasts for weeks, finally turning into draping hands of shiny pink berries. The shocking color and robust shape of the blossoms seems incongruous with the rest of the plant, adding to its mysterious appeal.

Plant Family
Melastomataceae

Other Common Name
Malaysian grape

Bloom Period and Seasonal Color
Rose pink flowers in early summer and fall

Mature Height × Spread
Up to 4 ft. × 3 ft.

When, Where, and How to Plant
The key to success with Malaysian orchid lies in the choice of its location and preparation of a suitable growing medium. Use one-third as much organic matter as other planting material, whether you are planting in existing garden soil or potting mix, to create a soil that is organic and fertile, yet well drained. A combination of ground bark and compost works well. Grow one Malaysian orchid in a standard 10-inch clay pot or space plants 18 inches apart in garden beds. Potted plants can be sunk into a bed of mulch for the summer, or planted outdoors in spring and repotted if necessary in late fall. Select a warm location in bright shade for beds and pots and mulch all plantings with an organic material other than pine straw.

Growing Tips
Potted Malaysian orchid plants can grow year-round in a bright indoor space away from direct sun, but they benefit from time outdoors in warm weather. The plants are even-tempered in nature and react poorly to extremes of drought or prolonged flooding. Water and fertilize very regularly in pots and beds. Use a complete balanced soluble fertilizer mixed at half strength once monthly all year. Work rotting mulch into the soil and replace it promptly with fresh mulch.

Care and Propagation
Be on the lookout for spider mites. Malaysian orchids can be propagated by division. See "Propagation" (page 36) for details.

Companion Planting and Design
Grow Malaysian orchids with rattlesnake plants, impatiens, and justicia for contrasts in colors and shapes. Use them to brighten up any shady space with an edging of caladiums and impatiens in front to repeat its rosy pinks.

Try These
M. magnifica is an epiphytic form of Malaysian orchid with foot-long leaves.

Nerve Plant
Fittonia albivenis

When, Where, and How to Plant

Choose a warm shady or mostly shady location outdoors to grow nerve plant as an annual or in pots to spend the summer. If the shade is slightly dappled with sun, try it, but remember to water more often. Provide a soil for beds and pots that is rich, organic, and well drained. Amend native soils and potting mixes with materials like compost or ground bark to enrich their organic content and improve drainage. Use one-third as much organic matter as other planting material. Grow one nerve plant in a 6-inch pot or space plants 6 inches apart. Take care that nerve plant will be growing at the same level it was in the pot; do not bury or elevate its center. Water well and mulch new plantings. As the mulch decomposes, work it into the soil.

Growing Tips

Nerve plants are moderate in nature and do not react well to extremes. Water and fertilize very regularly in pots and beds. Water, let the soil dry only slightly, and water again. Add a complete balanced soluble fertilizer mixed at half strength to the water once a month all year. Garden plants lift easily in the fall to be potted up to spend the winter indoors. Pots can grow year-round in a bright indoor space, but their growth will be thicker outdoors in summer.

Care and Propagation

Watch for slugs and snails on young plants and for mealybugs on new growth. Propagate by cuttings.

Companion Planting and Design

Let nerve plants be living mulch around Malaysian orchids and lady palms. Group them with peacock plants and Persian shield for a crazy quilt of patterned leaves, or pot them up with firecracker flowers and dumb canes.

Try These

Red-veined *Fittonia* tolerates slightly more sunlight and is often called mosaic plant.

Precious and eye-catching as it creeps around at ground level, nerve plant is native to the Andean rainforests of South America. Dark green and 4 inches long, the leaves would escape notice were it not for their incredible network of white veins crisscrossing their surfaces. Every curve and angle in the designer's notebook is covered in their patterns, proof to any who need it that all art has organic origins. It is as if a manic cake decorator were turned loose on the leaves to ice them white. Nerve plants, often listed by their previous Latin name F. verschaffeltii, grow no taller than half a foot; the roots are shallow, spread slowly, send up new leaves, and move on. Occasionally, greenish flowers are set among greenish bracts, but it seems they are aware of their stunning setting and sit quietly among the spectacular leaves.

Plant Family
Acanthaceae

Other Common Name
Mosaic plant

Bloom Period and Seasonal Color
Evergreen leaves with white and purple veins

Mature Height × Spread
Up to 6 in. × 6 in.

Nun's Orchid
Phaius tankervilleae

Like a siren call, pots of nun's orchids in bloom draw in the crowds at garden shows and nurseries. Instead of the reverent reference to the nodding flowers, this floozy of an orchid could be known as the painted lady or, to be kinder, the party girl. Her fragrant, tubular flowers are an avant-garde take on the classic orchid shape in a wacky array of colors. Bloom clusters open up the stem over months, as if the plant cannot stop itself. Like a teenager experimenting with makeup, each aspect of a nun's orchid is a distinct hue, unusual and striking when taken together. Hunky emerald green leaves are swords that shelter the fat stems and flamboyant flower spikes. It is the rosy throat that mesmerizes pollinators that seek the nun's nectar, but the burgundy brown sepals with white backs are just as irresistible to gardeners.

Plant Family
Orchidaceae

Other Common Name
Swamp orchid

Bloom Period and Seasonal Color
White, purple, and coral flowers in late spring

Mature Height × Spread
Up to 3 ft. × 2 ft.

When, Where, and How to Plant
Choose a warm, bright shady location outdoors to grow this plant as an annual or to place pots for the summer outside the tropics and warm subtropics. Indoors, a bright corner away from direct sunlight will help develop the early spring flower buds. Enrich existing garden soils and potting mixes with compost to increase their water-holding capacity, yet still retain their ability to drain. Consider a soaker hose in plantings with nun's orchids and plants with similar water needs. Water potted plants from below (using a saucer) to allow them to take up plenty. Space plants 18 to 24 inches apart in beds or provide a standard 10-inch plastic pot for each plant. Keep beds mulched with an organic material other than pine straw; as the mulch decomposes, work it into the bed or pot.

Growing Tips
These plants require a very regular watering schedule that allows their soil to barely dry out between irrigations. Use a complete balanced soluble fertilizer mixed at half strength once monthly all year. If grown in garden beds outside the tropics, the plants must be potted up for the winter. Nun's orchid is one of a few that can be propagated by laying the spent flower spike in a bed of damp sand and covering it halfway with the sand. In six weeks, baby plants should sprout.

Care and Propagation
You will not have to worry about pests, there are few to bother it. Propagate by division of tubers, offsets, and by burying the flower spikes in moist sand.

Companion Planting and Design
Group nun's orchids with ferns, Malaysian orchids, dumb canes, and justicia for deep greens even without flowers. Surround them with caladiums, impatiens, and begonias of all kinds.

Try These
P. microburst 'Wild Thing' is fancy, painted orange with maroon blush and a frilly cream-colored lip.

Patchouli
Pogostemon cablin

When, Where, and How to Plant
Indoors and outside, select a site that offers light but is shaded from direct sun all day. Patchouli is decidedly not drought tolerant and will die without consistently available water. The plants can grow in sunnier locales but they will need even more water to thrive. Prepare a garden soil or container growing mix that is richly organic but still drains well. Use ground bark, compost, and other organic matters to amend and enrich soils as needed. Plant clumps 18 inches apart in beds or provide a standard 10-inch pot for one patchouli plant. Be sure the new plants are growing at the same height they were originally. Mulch the plantings to a depth of 2 inches in beds, 1 inch in pots, with organic material.

Growing Tips
Use a soaker hose or reservoir pot to ensure enough water is available for patchouli plants. Fertilize monthly with a soluble, all-purpose garden formula. Patchouli can be pinched or cut back anytime its stems become lanky or to harvest them. To bring plants into flower, provide twelve hours of darkness for six weeks by putting pots in a dark closet or covering them outdoors with cardboard boxes. Dig up plants or root cuttings in the fall for wintering indoors.

Care and Propagation
There are few pests to "bug" patchouli. Propagate by cuttings.

Companion Planting and Design
Grow patchouli with African mask and striped blushing bromeliads for contrasting leaves and patterns. Surround them with ferns and prayer plants, or fill the space around them with nerve plants.

Try These
Many members of the Lamiaceae family are grown for their fragrance and oils, including mints and basils.

Native to India and other parts of tropical Asia, patchouli gets its name from Hindustan words for "green" and "leaf," hinting at the plant's most abundant asset. Compact, purplish red to brown stems create a rounded shape covered in puckered leaves with dark veins and lightly ruffled edges. They are almost olive green in color with eerie bronze overtones in some lights. It is not difficult to get positively drunk on the fragrance of patchouli, musky, rich, and as long-lasting in a closed room as the famously intoxicating gardenia or hyacinth. Patchouli oil became the signature fragrance of the 1960s counterculture but has been well known for centuries in Eastern cultures. In the world of aromatherapy, patchouli is reputed to bring harmonious energy to a space. The flowers have even more concentrated oils than the leaves, but they are only present as the days grow shorter in late summer and fall.

Plant Family
Lamiaceae

Other Common Name
Vicks plant

Bloom Period and Seasonal Color
White and purple in late fall

Mature Height × Spread
Up to 3 ft. × 2 ft.

Peacock Plant
Calathea makoyana

Peacock plant is a sight to behold, growing rather low to the ground with riotous leaf patterns that change subtly as the light passes over them. Each round leaf of a peacock plant begins its life at the top of a skinny stem, curled up tightly with only its pinkish red petticoat showing. As if the process was just too much, the leaf slowly unfurls to reveal its topside's green glory. They are delightfully elegant, with a light green background so rich that its dark green markings look raised, like an embossed invitation. Marked in a thickly barred pattern extending to feathery leaf edges, peacock plants seem to be hand-painted in sure strokes by a master. Dozens of stems rise from a clump at ground level to create a leafy explosion. Peacock plant leaves are relatively thin to the touch, but the broad, firm strokes in their markings give it a robust appearance.

Plant Family
Marantaceae

Other Common Name
Cathedral plant

Bloom Period and Seasonal Color
Patterned leaves in green, silver, and purple year-round

Mature Height × Spread
Up to 2 ft. × 1 ft.

When, Where, and How to Plant
Peacock plants grow best in bright shade. Direct sun will cause sunburned leaves, while very low light causes leaf colors to pale. A warm, bright, humid space indoors works well all year for peacock plants. Outside, raised beds of highly organic, well-drained soil accommodate them for months in warm weather. Amend garden soils and potting mixes by adding organic matter such as compost and ground bark to enhance both drainage and water-holding capacity. Space plants 1 foot apart in beds or give individual plants a pot 8 to 10 inches by 6 to 8 inches. Prepare to water outdoor plantings very regularly, using soaker hoses or drip irrigation. Raise humidity indoors by misting plants and grouping them together on beds of gravel kept moist.

Growing Tips
Put peacock plants on a frequent water regime but do not allow them to stand in water. Fertilize regularly year-round with any all-purpose formula. Keep plants mulched with 1 inch of organic matter such as ground bark. Work the mulch into the bed or pot as it rots and replace with fresh organic material. Move pots outdoors in the spring and summer or plant in beds. In the fall, reverse the process and pot up for the winter indoors.

Care and Propagation
Watch out for spider mites in hot weather. Propagate by separating.

Companion Planting and Design
This plant is excellent in combination with golden brush gingers and cane begonias for diverse, yet complementary leaf shapes, and looks sweet beneath Malaysian orchids and lady palms.

Try These
C. majestica has light green leaves with pinkish white stripes on top with purple beneath.

Persian Shield
Strobilanthes dyeranus

When, Where, and How to Plant
Select a location in bright shade outdoors to grow Persian shield as a summer annual or to sink pots of it into a deep mulch bed. Dappled sun sites can work if enough water is available. The plants are not drought tolerant, nor are they happy in wet soil. Prepare a soil for beds and pots that is richly organic, fertile, and drains well. Amend existing soils and potting mixes with organic matter like compost to enrich them and improve drainage. Use one-third as much organic matter as other planting material. Give each Persian shield a standard 8- to 10-inch pot or space plants 10 inches apart in garden beds. Mulch around the plants with an organic material to a depth of 2 inches. Use a mulch other than pine straw so it can be worked into the soil as it rots naturally.

Growing Tips
Persian shield reacts poorly to extremes of drought and flooding. Its leaves will turn pale and fall off in reaction to such stress. Use a complete balanced soluble fertilizer mixed at half strength once monthly all year. When they are grown in garden beds outside the tropics, Persian shield plants must be dug up and potted up for the winter. Give them a bright indoor space away from direct sun. Pinch plants to promote compact growth.

Care and Propagation
Keep your eyes open for slugs and snails on young plants. Propagate by cuttings anytime.

Companion Planting and Design
Persian shield is easily grown in a friendly group with bird's nest ferns, nun's orchids, caladiums, justicia, and peacock plants. Edge a stand of Persian shield with nerve plants.

Try These
S. anisophyllus, called goldfussia, is shrubby with thin, purple leaves and pink, bell-shaped flowers.

One clump of Persian shield puts a dazzling, iridescent exclamation point on any mixed planting. A group of them creates a stained-glass window in a garden with deep purple-green leaves painted in feathery patterns of rosy lavender with silver tones. The leaf surface and its dark veins shine through the lighter colors to create a show-stopping, glistening, faceted look. Native to Myanmar and other parts of tropical Asia, these plants bloom in flower spikes that are attractive, but not spectacular, and will not be missed if they are removed. (Their presence on a plant diverts its energy from the production of more leaves.) The stems are easy to pinch back and respond with thicker growth; those cuttings root easily in water or potting soil. Persian shield can tolerate full sun in cold climates, but elsewhere it is a colorful stalwart in the shade garden.

Plant Family
Acanthaceae

Other Common Name
None

Bloom Period and Seasonal Color
Purple, pink, and silver leaves year-round

Mature Height × Spread
Up to 4 ft. × 2 ft.; averages 2 ft. × 1 ft.

Philippine Orchid
Spathoglottis plicata

Native to tropical Asia and the Philippine Islands, this terrestrial orchid boasts fat clusters of flowers that attract hummingbirds and butterflies. The clusters open dramatically from the top down and can last three months. Each bud is puffy with fleshy petals and flower parts that pop open over a few hours to reveal a classic orchid-shaped flower. The brilliant blooms are clad in shades of pink, purple, lilac, and lavender, warm and inviting like pretty maids at a debutante ball. They stand on stems just above dense clumps of bright green foliage. The leaves are thuggish in defense of the maidens—thick, strappy and pleated like so many coy fans spread to shield them. Admittedly a commoner in the world of orchids, even the terrestrials, Philippine orchid deserves respect for its easy-growing nature, in beds and containers, indoors and out.

Plant Family
Orchidaceae

Other Common Name
Ground orchid

Bloom Period and Seasonal Color
Lilac-purple flowers appear sporadically year-round

Mature Height × Spread
Up to 2 ft. × 2 ft.

When, Where, and How to Plant
Philippine orchids grow best in soil that is fertile, organic, and well drained. Mix existing garden soils and potting mixes with compost, ground barks, and other aged products such as manure. The soil must be able to hold moisture and be watered regularly without producing a saturated root zone. Flowers are most abundant in full sun, but the plants adapt well to part sun and bright indoor areas if the sites are warm and humid. Choose a sheltered location away from strong winds and salt sprays for this orchid. Space plants 2 feet apart in beds or provide a pot 12 inches wide and deep for one clump. Do not bury the crown; plant at the same depth or slightly higher than they were growing originally.

Growing Tips
Soaker hoses can be the simplest, most reliable irrigation strategy for outdoor plantings of terrestrial orchids. Such water-lovers benefit from a timer on that water supply to deliver it regularly. Keep plantings mulched to a depth of 2 inches in beds, 1 inch in pots, and as organic mulch rots naturally, work it into the soil. Fertilize with a complete granular formula three or four times each year or use an equivalent soluble product every other month.

Care and Propagation
There are few pests to bother these plants. Propagate by division of the clumps. See "Propagation" for details.

Companion Planting and Design
Grow this orchid with reed stem orchids to display their diverse shapes. Group them with dwarf palms and lobster claw for rugged good looks, and fill a garden bed or container "floor" with star jasmines around them.

Try These
Although quite rare, the white Philippine ground orchid is worth pursuing.

Polka Dot Plant
Hypoestes phyllostachya

When, Where, and How to Plant
Provide a site to grow polka dot plants in garden beds or containers that is sunny, partly sunny, or in dappled sun. Sunnier sites will require more water but the growth will be more compact. Leaves can pale in full sun without adequate water. Prepare an organic, fertile, well-drained soil by amending existing garden soils and potting mixes with organic matter such as compost and ground bark. Space polka dot plants 8 to 12 inches apart in beds or large mixed pots or give them a standard 6-inch pot for each plant. These plants thrive in humid, hot weather outdoors in beds and adapt well to winter as potted plants indoors where they are not hardy. Water well after planting and mulch around each plant with 2 inches of organic material.

Growing Tips
Water and fertilize polka dot plants often enough to allow the plants to dry out just slightly between irrigations. Pinch young plants to encourage bushy form and cut back older plants if they become leggy. Use a general-purpose fertilizer, such as a soluble formula mixed in water, regularly year-round. Alternatively, use a slow-release formula added to the soil four times a year. Mulch outdoor plants well in the fall but do not allow them to become waterlogged.

Care and Propagation
Be on watch for slugs and snails. To grow more, propagate by cuttings.

Companion Planting and Design
Combine with impatiens and begonias in a shady site, or with chenille plants and coleus in sunny spaces. Plant an edge of polka dot plants in front of tall vines such as hyacinth bean and moonflowers.

Try These
'Freckle Face' has green leaves with white veins and 'Pink Splash' has pink leaves with green veins.

Native to Madagascar, polka dot plants display vibrant colors—shiny green leaves marked with neon pink veins in irregular, freckled patterns. Like designer shoes, the small plants draw the eye down instantly and keep the focus squarely at its feet. Each leaf is flecked, mottled, and painted with rosy pink as if its colors had been applied in a spin art toy. The leaves are fat daggers that pop out in pairs from compact stems to form mounds of impressionistic color near ground level. Blue and white flower spikes emerge from the clumps in bright clusters, but their presence diverts energy needed to grow more leaves. Polka dot plants can be returning perennials in the subtropics, where the first emerging leaves are considered a sign of spring. Their sumptuous leaves provide living mulch for larger plants while growing into a firm edge to underline the scene's tropical strengths.

Plant Family
Acanthaceae

Other Common Name
Freckle face

Bloom Period and Seasonal Color
Green or pink leaves with pink or white mottles and veins, inconspicuous blue flowers in summer.

Mature Height × Spread
Up to 1 ft. × 1 ft.

Prayer Plant
Maranta spp.

Native to South America, this is a group of plants that share a bizarre trait: a natural trigger at the base of each leaf closes the leaf in response to darkness. The sight of clasped leaves upraised inspired its common name, prayer plant. Within the group, three leaf patterns dominate in the trade and all are called prayer plant. Dark green, oval leaves begin life tightly rolled, with only their reddish purple undersides showing. As they unfurl, the jewel tones of the upper surface reveal their oddball patterns. One pattern paints the midrib in a whitish-pale green, fat feather created by the leaf veins with dark splotches in between. Another adds red to the leaf veins to produce a classier, less thuggish look. The third prayer plant goes to extremes, its feather patterns extending to the leaf edges. Versatile prayer plant puts color and motion into damp, shady spaces with easy grace.

Plant Family
Marantaceae

Other Common Name
Praying hands

Bloom Period and Seasonal Color
Green and purple leaves, white flowers occasionally

Mature Height × Spread
Up to 1 ft. × 3 ft.

When, Where, and How to Plant
Prayer plants can grow outdoors year-round in frost-free areas and as a shade garden plant in warm weather elsewhere. They are easygoing potted plants that can be sunk into a bed of mulch for the summer outdoors and grown indoors. These plants thrive in fertile, organic, well-drained soil amended with compost, ground barks, and other aged products such as manure. Whether in a garden bed or a container, the soil must be able to hold its moisture and tolerate watering regularly without causing the plants to stand in water. Space them 12 inches apart in beds, or cluster three prayer plants in a 12-inch pot that is wider than it is tall. Water new plantings well and mulch beds and pots with 2 inches of organic material.

Growing Tips
To provide a very reliable source of water for prayer plants, consider soaker hoses or container drip systems on timers to make this task foolproof. Prayer plants growing in semi-shade or dappled sun will need considerably more water than those in shade. Fertilize regularly all year with a soluble, general-purpose formula. As the mulch rots, work it in to the soil for added nutrition and replace it promptly.

Care and Propagation
Watch out for mealybugs attacking prayer plants. Propagate them by dividing the clump.

Companion Planting and Design
Start a collection of shade-loving foliage by pairing with jade vines and umbrella plants. Use it alongside sanchezias, in front of lobster claw and reed stem orchids, or as a groundcover with vines like flaming glorybowers.

Try These
Calatheas such as *C. zebrine* are related to prayer plants but have leaves that are long and narrow. *Maranta arundinacae* has no variegation and is grown as the starch crop arrowroot.

Queen of the Night
Epiphyllum oxypetalum

When, Where, and How to Plant

These plants can grow for years in frost-free, dappled sun garden beds that have excellent drainage. Where they are not hardy, pots can spend summers outdoors and then move happily indoors to a bright room for the winter. They will grow slightly faster in brighter light but cannot tolerate direct sun, indoors or outside. Prepare an organic, well-drained soil by amending garden soil and potting mixes with compost or ground bark to improve its structure. Plant one queen of the night in a standard 10-inch pot or space plants 1 foot apart in beds. Nestle the plants into the soil for stability because they grow heavy with leaves that break easily if the pot is tipped over. Mulch pots and beds lightly with organic material such as ground bark.

Growing Tips

Overwatering can cause rot at the roots. Let pots and plantings dry out between waterings. The plants cannot tolerate wet feet, but neither are they drought tolerant for very long. Fertilize using a soluble, general-purpose formula as directed on the product label at full strength in the spring and summer, then at half strength in the fall and winter. Queen of the night will produce more leaves and blooms if it's repotted only when roots are constricted.

Care and Propagation

Watch for beetles on any new growth. Propagate by stem cuttings and seeds.

Companion Planting and Design

This flowering cactus can bloom after bush lilies and earth star in a bold show. Jade plants make excellent companions indoors and on the deck. Plant with schefflera and peperomias for stunning green contrasts.

Try These

Though not all are night blooming, scores of other flowering cacti are as easy to grow.

Call the neighbors, put on a pot of coffee, and wake the children—it's midnight and queen of the night is opening its flowers! The spectacular white flowers as big as softballs open around midnight and are long gone by noon. The ritual repeats itself for several nights and recurs throughout warm weather as flowers appear in flushes. Long and fat, the flower buds swell for days and finally open, exquisite and complex layers of petals exploding with fancy stamens that put spider lilies to shame. This flowering cactus sports long, scalloped leaves that zigzag crazily in every direction; they break and root easily. The buds appear at branch ends and in the leaf scallops in great numbers, ready to bloom, be pollinated, and turn into shiny magenta footballs holding seeds that are black, poppyseed-looking things packed tight for the next generation.

Plant Family
Cactaceae

Other Common Name
Night-blooming cereus

Bloom Period and Seasonal Color
White flowers in the summer followed by magenta seedpods

Mature Height × Spread
Up to 3 ft. sprawling in every direction

Rattlesnake Plant
Calathea lancifolia

Rich jewel tones and distinctive leaf patterns mark rattlesnake plant, the quintessential ambassador for tropical plants. Unforgettable to see, easy to grow, and forgiving of most human frailties, its leaves are stiff, sometimes wavy swords with amazing two-sided interest that rise from a round, flat rosette. Each leaf barely overlaps its neighbors; instead, they emerge from all over the rosette and stand up at angles to each other. The result of this exotic arrangement is a rare view of both leaf sides at once from one vantage point. One expects that plants native to Brazil will be colorful, and rattlesnake plant never disappoints. The top surface of its leaf is light green, silver, or occasionally yellowish with very regular patterns of darker dots and bars that really do resemble a snake's back. Not content to be blah, a leaf's underside shines in deep burgundy red, creating contrast to the upperside as only a tropical can do.

Plant Family
Marantaceae

Other Common Name
Rattlesnake calathea

Bloom Period and Seasonal Color
Prized year-round for leaves that are green with purple spots above and red-purple below

Mature Height × Spread
Up to 2 ft. × 1 ft.

When, Where, and How to Plant
Indoors, any warm, bright room can sustain rattlesnake plants over the winter or year-round. Choose a site in shade or part shade for garden beds or pots summering outdoors. Direct sunlight, indoors or out, will rapidly dry the leaves and fade their colors. Prepare a soil for beds and containers that is rich in organic matter, fertile, and well drained. Amend garden soils and potting mixes with ground bark and compost in nearly equal amounts to ensure these qualities. Space rattlesnake plants 1 foot apart in beds and large planters or put one in a container 8 to 10 inches wide and 6 to 8 inches deep. Repot when water rushes through without percolating into the soil, indicating it has become rootbound, or prune the roots and repot in fresh growing mix.

Growing Tips
Water deeply and often, but do not allow rattlesnake plants to stand in water. Fertilize regularly with an all-purpose formula to keep new leaves coming from the rosette. Well-fed plants retain strong leaf color for months. Keep plants mulched with organic matter and work decomposed materials into the soil below. Clip off flowers when they appear to prevent competition with the leaves for water and nutrients. If clay pots dry out between waterings, repot into plastic ones for easier water management.

Care and Propagation
Keep an eye out for spider mites in hot weather. Propagate rattlesnake plants by separating the leafy clumps.

Companion Planting and Design
This plant is an ideal indoor garden plant and excellent paired with bird's nest ferns and Persian shield that echo its colors, with chocolate plants for textural contrasts, or alongside its cousin, peacock plants.

Try These
Also called rattlesnake plant, *C. crotalifera* is larger and slightly more cold hardy.

Red Abyssinian Banana

Ensete ventricosum

When, Where, and How to Plant

Select a space in full sun with a reliable irrigation source nearby. This is a large plant that grows well in garden beds and large containers but should be spaced for good air circulation in humid, rainy climates. Prepare soil that is richly organic, fertile, and well drained. Amend garden soil or potting mix with organic matter to achieve these conditions. Put one red Abyssinian banana plant in a 5-gallon nursery pot or plant it in a bed with 3 to 4 square feet to itself. Add ¼ cup composted manure to the planting hole and cover with 1 inch of soil. Plant the banana at the same level it was growing originally and slope the soil away from the stem. Mulch beds and pots with 1 to 2 inches of organic material.

Growing Tips

Water and fertilize regularly and let the plants dry out only slightly. Use a general-purpose garden fertilizer regularly from the spring through fall. Maintain organic mulch around the plants and work it into the soil when it rots. Dig up plants in the fall, root prune, and pot up. Remove the lower leaves to ease its transition indoors. It can be wintered in a cool, but not freezing, shed.

Care and Propagation

Watch for slugs and snails on young plants and chewing caterpillars attacking emerging leaves. Propagate by dividing a clump.

Companion Planting and Design

This is a centerpiece plant. Add to its drama with frangipani and bird of paradise. Surround it with Persian shield, pink summer snapdragons, and mandevilla vines for a complementary color scheme.

Try These

'Little Prince' and 'Truly Tiny' are small banana plants at 2 to 3 feet tall. The latter has variegated leaves. 'Maurelii' delivers huge red leaves.

Native to tropical Africa and Asia, red Abyssinian banana paints deep patterns on big leaves that open in spring and keep on coming with big boy leaf textures and muscular form in the summer garden. Like all banana plants, this one grows rapidly, unfurling huge leaves in a loose spiral around a thick stalk. Red Abyssinian banana, especially the cultivar 'Maurelii', outdoes other banana plants, the virtuoso whose solo is heartfelt and memorable. Each leaf becomes a symphony of refracted colors when the sun shines. Their red undersides burn through to the green upper leaves in hues of coppery red and feathery patterns that catch the light like a red-haired child's curls. The plants can bloom and make bananas that are not edible, but they do not make suckers at their base. The graphic and soulful presence of red Abyssinian banana more than makes up for these shortcomings.

Plant Family
Musaceae

Other Common Name
None

Bloom Period and Seasonal Color
Green-red and very red leaves year-round in warm environments

Mature Height × Spread
Up to 20 ft. × 10 ft.; average 8 ft. × 4 ft.

Red Ginger
Alpinia purpurata

Red ginger is a romantic plant, conjuring up leafy tropical groves where fantastic flowers wait to be discovered among dense stands of leafy stalks. This ginger lives up to that image in its native home on the Malaysian peninsula in the South Pacific, and it brings the same allure to gardens worldwide. Even without flowers present, red ginger delivers tropical pizzazz anywhere. Reedy stems shoot up and leaves unfurl in one dimension, like some corn plants. This single plane arrangement makes for a vigorous upright effect, as if each leaf is reaching to be top on the totem pole. The leaves fade from view quickly, however, when the flowers bloom. Dozens of rich red, silky bracts shield tiny white flowers, creating a curvy cone of astounding color depth. The power of red as an attractant gives the bracts a sure lure for insect pollinators, as effective as a bullfighter's cape.

Plant Family
Zingiberacae

Other Common Name
Ostrich plume

Bloom Period and Seasonal Color
Showy red or pink flowers in the summer and fall; year-round in the tropics

Mature Height × Spread
Up to 12 ft. × 4 ft. in the tropics; up to 6 ft. × 3 ft. as an annual or potted plant

When, Where, and How to Plant
Red ginger thrives in fertile, organic, well-drained soil amended with compost, ground barks, and other aged products such as manure. Whether in a garden bed or a container, the soil *must* be able to hold its moisture and tolerate watering regularly without causing the plants to stand in water. Flowers appear in the second year and are more abundant in a sunny site, but this ginger adapts well to partly sunny areas and bright indoor areas. Choose a sheltered location, away from strong winds and salt sprays. Plant young red gingers close together and at the same depth they were growing originally. Space them 18 inches apart in beds, or cluster three plants in a 10-gallon pot. Group red gingers with other water-lovers to simplify maintenance.

Growing Tips
Use soaker hoses or another reliable of irrigation for outdoor plantings and consider a timer to deliver it regularly. Keep plantings mulched to a depth of 2 inches in beds, 1 inch in pots. Fertilize red gingers with a complete granular formula three or four times each year; as organic mulch rots naturally, work it into the soil for additional nutrition. Crowded stems may dry out or rot and slow flowering or stop it altogether in mature plants.

Care and Propagation
These plants suffer from few pests, but watch for mealybugs. Propagate red ginger by offshoots and roots. Its seeds are viable but slow to germinate.

Companion Planting and Design
Plant a bed of red and pink gingers for cut flowers, or let them shine alongside water-friendly companions like banana and rice paper plants, butterfly flowers, and sanchezias.

Try These
Look for 'King Red' and 'Queen Pink' for superb, long-lasting flowers.

Reed Stem Orchid

Epidendrum spp.

When, Where, and How to Plant

Reed stem orchid thrives in fertile, organic, well-drained soil amended with compost, ground barks, and other aged products such as composted manure. Whether in a garden bed or a container, the soil must be able to hold its moisture without causing the root zone to stand in water. Choose a sheltered location that is sunny, or partly so, and away from strong winds and salt sprays for these orchids. Select a mix of colors for the best effect. Plant young plants close together at the same depth they were growing originally. Space the plants 2 feet apart in beds, or cluster three plants in a container 12 inches wide and deep. Mulch around plants with 1 to 2 inches of organic mulch.

Growing Tips

Use soaker hoses or another reliable irrigation source for reed stem orchids and consider putting a timer on that water supply to deliver it regularly. Keep plantings mulched to a depth of 1 to 2 inches in beds, 1 inch in pots, and work decomposed mulch into the soil. Fertilize with a complete granular formula three or four times each year. Crowded stems may dry out or rot and slow flowering or stop it altogether in mature plants.

Care and Propagation

There are few pests to bother reed stem orchids (meaning few to bother you!). Propagate by dividing a clump. See "Propagation" (page 36) for details.

Companion Planting and Design

Grow reed stem orchids in a cluster in front of ruffle ferns and umbrella plants. Group them with water-lovers like fragrant heliotrope, Philippine orchids, and passion flowers, or let star jasmines crawl around at its feet.

Try These

E. radicans is a terrestrial reed stem orchid and *E. pseudepidendrum* has stunning orange flowers.

Native to the American tropics, reed stem orchids are a hugely diverse family of true orchids that are popular worldwide for their sultry flowers with pouty, painted lips. The genus name, Epidendrum, *means "on tree" in Greek; an epiphyte attaches itself to tree bark for support and relies on available rainfall and sunlight to survive. Such dependence on the kindness of strangers motivates a plant to evolve and adapt to a wide range of habitats. There are reed stem orchids native from jungles to dry savannas and cloud forests, creating great diversity in plant size, leaves, and flowers. Some are even terrestrial, despite their name. All have reedy stems, but some are upright, while others creep around the soil surface. The leaves are green and thick, but their numbers vary widely from plant to plant. The flower spike is a peduncle holding up a flat pinwheel of waxy petals that become a tablecloth for the centerpiece of true, vampy flowers.*

Plant Family
Orchidaceae

Other Common Name
Crucifix orchid

Bloom Period and Seasonal Color
Orange, red, yellow, white, or purple flowers year-round

Mature Height × Spread
Up to 4 ft. × 4 ft.

Rex Begonia
Begonia rex

Gardeners are often jealous of those who grow Rex begonias successfully. The plants look so complicated that they must be hard to grow. They are not. Rex may ask a bit more than some plants, but they are unbelievably rewarding. Exotic, surprisingly intricate leaf patterns arch stiffly on fleshy stems growing from shallow rhizomes. The leaves are bold, thick, and sometimes textured, adding to their tropical impact. Some have a tightly curled center on each leaf that gives them a deeply patterned, three-dimensional look. Others are shaped like fat teardrops and painted in rich shades of green, red, silver, pink, and purple. The original Rex is said to have been native to northeast India and was imported unintentionally to England in the 1850s. Fans of Rex begonias have that serendipity to thank for initiating a breeding frenzy that resulted in even more striking leaves.

Plant Family
Begoniaceae

Other Common Name
Painted leaf begonia

Bloom Period and Seasonal Color
Prized for intricate, colorful leaf patterns year-round

Mature Height × Spread
Up to 18 in. × 18 in.

When, Where, and How to Plant
Grow in soil that is richly organic and well drained. While Rex is usually grown in containers outside the tropics, good garden soil makes a fine summer home. Amend garden soils and potting mixes to ensure they can keep its root zone hydrated yet allow water to drain away rapidly. Test a pot of the mixed soil by watering it well. If water does not drain promptly, add more ground bark. Grow in shade or part shade outdoors away from wind and salt spray. Plant each Rex in a pot 8 to 10 inches wide and deep or space plants 1 foot apart in beds for ample root space and good air circulation around the leaves. Dig up garden plants and grow them in pots indoors during the winter in a brightly lit, humid space.

Growing Tips
Rex begonias are the most finicky in the begonia family and do not adapt to change readily. Repot only when water rushes through the container, indicating a dense rootball that cannot absorb it. Their needs are specific but not difficult to provide. With regular water and fertilizer applications, new leaves will emerge year-round. Rex enjoys more humidity than other begonias, but like them, plants need to dry out just slightly between waterings.

Care and Propagation
Watch for mealybugs deep in the leaf axils and for leaf spot. You'll probably want more, so propagate them by leaf and rhizome cuttings.

Companion Planting and Design
Be mindful of Rex's preference for high humidity when choosing indoor sites. Group pots together or with companions like Persian shield, flamingo flowers, and lady palms.

Try These
'Annie' has sharply cut leaves in pink and silver and 'Curly Lion' is gold and red.

Sanchezia

Sanchezia speciosa

When, Where, and How to Plant

Sanchezia shows off its brightest colors in full sun but will be quite colorful in part sun, where it will need somewhat less water. As sunlight decreases, green coloration increases. These plants thrive in fertile, organic, well-drained soil amended with compost, ground barks, and other aged products such as manure. Whether in a garden bed or a container, the soil must be able to hold its moisture and tolerate watering regularly yet not be flooded. Space the plants 1 foot apart in beds, or cluster two plants in a 10-gallon pot. Be sure to plant sanchezia at the same depth they were originally. Plants can be dug up and potted in the fall outside the tropics, or can be carried over the winter as rooted cuttings.

Growing Tips

When growing sanchezias in full sun, especially in pots, very regular watering will be needed. Soaker hoses are ideal in bed plantings and, if necessary, can be put on a timer for your convenience. Fertilize with a complete formula as often as directed on the label in the spring and summer, and less often in the fall and winter. Keep plantings mulched to a depth of 2 inches in beds, 1 inch in pots; as organic mulch rots naturally, work it into the soil.

Care and Propagation

Lucky for the gardener, sanchezias suffer from few pests. Propagate by rooting cuttings in the summer. See "Propagation" (page 36) for details.

Companion Planting and Design

Combine with red gingers, calico plants, and lobster claw for great contrasts. Surround sanchezias with star jasmines in a big pot or grow them with fragrant heliotrope for stunning color combinations.

Try These

Other sanchezias are seldom found outside their native ranges and botanic gardens.

Brightly striped and crisp, sanchezia takes charge of any scene like an officer in uniform, quickly asserting the power of strong lines to calm a crowd. Such patterns are comforting because they are regular and predictable, unlike the impressionistic painted leaves of many tropical plants. Those who find crotons and copperleaf plants too busy for their style will embrace the leaf patterns in sanchezia. Thick and pointed at their tips, the leaves use dark green and golden yellow to present an assortment of stripes that vary in size but are pleasantly consistent. The plants are thick, with stems and leaves that overlap to create rounded shapes. These plants are staples of tropical landscapes, can become returning perennials in the warm subtropics, and are excellent container plants in any climate. Native to Ecuador and Peru, sanchezia is named for Joseph Sanchez, a nineteenth-century botany professor from Cadiz, Spain.

Plant Family
Acanthaceae

Other Common Name
None

Bloom Period and Seasonal Color
Yellow and green leaves year-round

Mature Height × Spread
Up to 6 ft. × 3 ft.; average 3 ft. × 2 ft.

Scarlet Sage

Salvia coccinea

What's in a name? In this case, a degree of confusion since two plants go by the same name. The popular bedding plant Salvia splendens is also called scarlet sage, but it is less desirable because most of its hybrids do not attract pollinators or hummingbirds, are less hardy, and seldom reseed. The plant to grow is S. coccinea, a hummingbird magnet of the first degree. Native to the American tropics, it has naturalized everywhere freezing temperatures are brief or nonexistent. Its tall stems are square and show off triangular leaves that unfurl opposite each other along its length. One-inch-long, glowing red flowers spin out around the stems in a freewheeling, airy arrangement. Each blossom is a tube that flares open with prominent yellow stamens atop petals that are elongated and scalloped, a natural neon sign. Scarlet sage is easy to grow and never a bully, and is a returning perennial in much of the subtropics.

Plant Family
Lamiaceae

Other Common Name
Texas sage

Bloom Period and Seasonal Color
Red flowers in the spring through fall, sporadically all year

Mature Height × Spread
Up to 2 ft. × 1 ft.

When, Where, and How to Plant
Scarlet sage is easy to grow as an annual or perennial in garden beds in the tropics and subtropics, and as a container plant everywhere. Choose a site in sun or part sun near a reliable water source and be prepared to use it. Wilted sages will stop growing and fail to bloom. Prepare a soil in garden beds or containers that is organic, fertile, and well drained. Space plants 8 to 10 inches apart in garden beds and mixed planters, or plant three small plants in a standard 10-inch pot. Scarlet sages flourish in garden soil in warm weather, and can be dug up and potted in the fall in the temperate zone. Mulch beds and pots with 1 to 2 inches of organic material.

Growing Tips
Water scarlet sage plants regularly, and let them dry out only slightly between irrigations. Use a general-purpose garden fertilizer as often as directed from the spring through fall and less often in the winter. Maintain organic mulch around the plants and work it into the soil as it rots. Keep old flowers clipped off of scarlet sage until the fall to prevent seed set. Cut back the plants if they stop blooming in very hot weather.

Care and Propagation
Watch out for slugs and snails. You can propagate scarlet sage by seed and cuttings.

Companion Planting and Design
Fill a pot with scarlet sages and bat face cupheas for a small space tropical touch. Use them as filler plants or edging with chenille plants, copperleafs, and Chinese hat plants for a color palette in shades from red to bronze.

Try These
Salvia is a huge genus that includes Mexican bush sage, pineapple sage, and *many* perennial plants.

Silver Vase Bromeliad
Aechmea fasciata

When, Where, and How to Plant

A mature silver vase bromeliad produces offsets (pups) around its base after flowering. Harvest (or purchase) 3- to 4-inch offsets. To remove an offset, remove the mother plant from its pot and brush away the soil to expose the connection. Gently separate the two. Let the young plant form a callus on its base, then plant in standard clay pots at least 6 inches across or wire them onto boards to hang outdoors. Remember, these are epiphytes that need support. Prepare a potting mix that is very well drained, such as a lightweight potting mix and sand. Put the pots or boards in shade for two weeks and water lightly but do not let the pups dry out. Once established, silver vase grows well in shade or dappled sun.

Growing Tips

Water board-mounted plants daily and do not allow the wires to cut into the leaves. Keep the soil moist but not saturated. If the soil stays wet, repot the plant immediately. However, maintain 1 inch of water inside the vase itself. Add soluble fertilizer to the water every month, all year long. Pale leaf color or failure to bloom after three years indicates a need for more sun.

Care and Propagation

Few, if any, pests bother silver vase bromeliads but watch out for bromeliad scale. Empty vases occasionally to prevent mosquitoes from breeding and to prevent rot. Propagate them from pups growing at the base of a plant.

Companion Planting and Design

Group them with equally striking plants with similar needs such as bamboo orchids, African gardenias, and rose cactus. Elevate a pot of silver vase bromeliads on a pedestal amid a bed of Persian shield and licorice plants. It is striking!

Try These

'Silver King' has the grayest leaves while 'Primera' has greener leaves and 'Variegata' has leaves with broad yellow stripes.

Once upon a time, silver vase was about the only bromeliad people saw outside the tropics. Its stunning leaves are cross-striped green and gray with a dusty coating that looks surreal, almost unearthly, at first glance. Silver vase bromeliad leaves are stiff and wide, and arch away from the stem to create a funnel or urn shape. One huge, pure pink bloom spike explodes from that vase, with a complex cone on top dotted with vivid purple flowers. The spectacular show lasts for weeks, but the plant is already busy making offsets to ensure the next act. In their native Brazil, silver vase bromeliads grow without soil, their roots holding fast to tree bark in the rainforest. Fortunately, they are equally at home in containers to be grown and enjoyed everywhere. Silver vase bromeliads in bloom are an unforgettable sight and inspire many to begin collecting the dramatic, easy-to-grow bromeliads.

Plant Family
Bromeliaceae

Other Common Name
Urn plant

Bloom Period and Seasonal Color
Pink flower spikes with purple flowers appear on three-year-old plants and last for weeks

Mature Height × Spread
Up to 2 ft. × 2 ft. with 3 ft. flower spike

Sky Flower
Plumbago auriculata

Native to South Africa, sky flower covers its light green leaves and stems with an abundance of sprightly azure blue blossoms. It is a rare color in the tropical color palette and as welcome as a cool breeze in August. Planted alone as a focal point or combined with the many yellows, reds, and oranges in the tropical garden, sky flowers deliver cooling contrast to every design. Sky flower will take on almost any shape desired with pruning to bring on its springy, viney stems. Left to its own devices, it sprawls and arches grandly, but sky flower can also be attached to a wall or keep a neat habit in a big pot with regular pruning. Bloom clusters form at the stem ends, fat heads of shiny blue pinwheels that resemble those of phlox plants, but, like everything in the tropics, sky flowers are bigger, bolder, and more beautiful.

Plant Family
Plumbaginaceae

Other Common Name
Cape leadwort

Bloom Period and Seasonal Color
Blue flowers late spring through the fall

Mature Height × Spread
Up to 6 ft. × 3 ft.

When, Where, and How to Plant
Choose a sunny site with a reliable water source nearby to grow sky flowers. The plants will stop growing and blooming without ample irrigation but yet they cannot tolerate wet feet. Prepare a soil in garden beds or containers that is both fertile and well drained. Amend existing garden soil and bagged potting mixes with sand and organic matters if needed. Space plants 1 foot apart in garden beds and mixed planters, or group two small plants in a standard 12-inch pot. Sky flowers flourish in garden soil in warm weather, can become perennial in the subtropics, and are not difficult to maintain over the winter as potted plants in the temperate zone. Mulch beds and pots with 1 to 2 inches of organic material such as ground bark.

Growing Tips
Water sky flower plants regularly and let them dry out only slightly between irrigations. Use a general-purpose garden fertilizer as often as directed from the spring through fall and less often in the winter. Cut plants back by half in the fall for garden plants and in late winter for those in containers. Established plants may have a reduced need for water and fertilizer but if they become woody, prune to stimulate new growth.

Care and Propagation
Sky flowers are bothered by few pests. Propagate them by taking tip cuttings in the summer and by seed. See "Propagation" (page 36) for details.

Companion Planting and Design
Combine sky flowers with yellow cestrums and night-blooming jasmines for contrasting leaves and flowers. Grow them in front of Chinese violets and surround them with polka dot plants, scarlet sages, and bat face cupheas.

Try These
'Royal Cape' has bright, cobalt blue flowers and 'Alba' is pure white.

Snake Plant
Sansevieria trifasciata

When, Where, and How to Plant

Snake plant thrives outdoors all year where it is hardy in shade, part shade, and even sunny sites if it is given time to acclimate. It can also be grown as a container plant year-round everywhere. Pots can spend the summers sunk into mulch beds in shade or dappled shade along with other, compatible plants. Plants can be installed directly into those same beds for the warm months, dug up in the fall, and repotted. Provide an organic, well-drained soil by amending garden soil and potting mixes with organic matters. Snake plants are best grown in clumps, but single plants in small pots tend to multiply fairly rapidly. Upright clumps can be planted close together. Be sure to plant at the same level the snake plant was growing originally. Mulch plantings lightly in beds and not at all in pots.

Growing Tips

Overwatering can cause this plant's leaves to collapse and fall over the side of the pot. Allow pots or plantings to dry out before watering again. Fertilize monthly with a soluble, general-purpose formula as directed in the spring and summer, then mixed at half strength in the fall and winter. Snake plants in containers should be repotted or separated when their pot is full of leaves or when water rushes through the soil, indicating it is rootbound.

Care and Propagation

Few pests bother this plant. You can propagate by rooting leaf cuttings, dividing plants, and potting up offsets.

Companion Planting and Design

Snake plants get along well with jade plants, dwarf schefflera, and blunt-leaf peperomias to create a diverse design in low light situations. Grow them with earth stars for an otherworldly container effect, or with crotons for contrasting shapes and colors.

Try These

'Moonshine' has fat silver-green leaves, *S. cylindrica* has cylindrical leaves, and 'Bird's Nest' is squat.

Native to South Africa, Zaire, and Zimbabwe, snake plant has an angry, aloof air. With leaves shaped like swords and daggers, it looks dangerous and off-putting. The good news is that this plant needs little attention and thrives on benign neglect. Crazy, wavy, cross-striped leaves, often with bold solid edges, are the green plant version of a snakeskin and give this plant its common name. Sharply pointed, thick, and unyielding, the leaves of the species live for years and give rise to their other common name of mother-in-law's tongue. Seemingly endless variations on the snake plant offer tall, upright, short, squat, and truly twisted leaf arrangements. Every color in the yellow-green-gray range is displayed in the snake plant group, with shades from darkest forest green to matte green, gray-green, and almost silver with whites and yellows so bright they glow. No tropical plant is easier to grow.

Plant Family
Asparagaceae

Other Common Name
Mother-in-law's tongue

Bloom Period and Seasonal Color
Dark green, gray, and yellow evergreen leaves

Mature Height × Spread
Up to 6 ft. × 2 ft.; averages 3 ft. × 1 ft.

Staghorn Fern
Platycerium spp.

Going to extremes with strange leaf shapes and odd-ball habits is the tropical way, and staghorn fern brings the X factor to these green games. If the category covers big, weird-looking plants that hang from trees with little apparent support, staghorn ferns win hands down. The family is native to habitats in Asia, Australia, and South America but plants are always recognizable for their distinctly different fertile and infertile leaves. The light green infertile ones are shaped like and known as "shields," while the smaller, often gray-green, fertile leaves are suspended below. The big shields are flat, scalloped, and send roots into the bark or growing mix to anchor the plants. The fertile leaves are dazzling, suspended like a rounded fork of fuzzy fingers. They are the source of the fern's spores, golden brown and arranged like a dot-to-dot puzzle or the horns of a stag.

Plant Family
Polypodiaceae

Other Common Name
Antelope ears

Bloom Period and Seasonal Color
Evergreen leaves

Mature Height × Spread
Up to 6 ft. × 8 ft. where hardy; averages 2 ft. × 4 ft.

When, Where, and How to Plant
In the tropics, staghorn ferns grow year-round attached to the bark of big trees. Elsewhere, they can be grown in pots of ground pine bark but are most often seen mounted on boards to be hung in trees during warm weather. Wrap the roots in chicken wire or a similar material containing damp potting mix and sphagnum moss. Use wire to secure the plants onto a piece of wood without cutting into the ferns. One staghorn fern can grow for years on 1 square foot of wood or board or in a container that is 12 inches deep and wide. In Nature, daily rain hydrates the plants but drips right off, and their needs are the same in the garden. Choose a site in shade or soft light that is near a reliable water source.

Growing Tips
Water very regularly to grow and maintain staghorn ferns. Use a soluble fertilizer occasionally mixed into the irrigation water and as a foliar spray. Keep the pockets around the shields free of debris, and sprinkle in a little compost a few times each year to ensure their nutrition needs are met. When staghorn ferns multiply beyond their supports, cut them apart and remount the individual plants.

Care and Propagation
Few pests other than scale insects bother staghorn ferns. Ants may infest the plants at times. Propagate by division.

Companion Planting and Design
Hang staghorn ferns amid golden brush gingers, split-leaf philodendrons, and elephant ears for gigantic tropical tone. Grow them alongside white bat flowers and striped blushing bromeliads for super contrast and colorful accents.

Try These
P. andinum, the South American staghorn fern, is a handsome, smaller staghorn with no cold tolerance.

Striped Blushing Bromeliad

Neoregelia carolinae

When, Where, and How to Plant

Choose a site outdoors where plenty of light is available but the plants will be shaded from direct sun all day to grow this bromeliad. A tabletop near a bright window amid a group of other plants will provide ample light and humidity inside a home or office. The plants may *seem* to be drought tolerant, keeping their color even if drought stressed, but they are not. Prepare a garden soil or container growing mix with plenty of ground bark, compost, and other organic matter added. The resulting growing medium should be very organic and fertile yet drain well enough to prevent puddling. Space plants 18 inches apart in beds or provide a standard 8- to 10-inch clay pot for each plant.

Growing Tips

Keep water in the cups of these bromeliads constantly and water the soil around the plants when it feels slightly dry. Rinse the leaves occasionally in dry weather. Add a soluble general-purpose fertilizer both to the fertilizing water and to the water in the cups once a month, alternating at half strength every other month. Repot when pots become congested. When flowers dry up, pot up the offshoots that will appear around the base of the mother plant. Let the "pups" rest before potting so their base forms a callus.

Care and Propagation

There are few pests of note except bromeliad scale. Empty vases occasionally to prevent vase rot and to prevent mosquitoes from breeding. Propagate by removing the offsets from around the mother plant.

Companion Planting and Design

Grow striped blushing bromeliads with white bat flowers and African mask for strong, exotic shapes, flowers, and patterns. Let them be an accent with elephant ears and prayer plants.

Try These

'Raphael' has very dark green leaves edged in nearly white with a wine-red rosette at its center.

With edges so sharp they look honed, in bright clear colors with flowers that last for months, plants in the Neoregelia *genus evoke visions of paradise all by themselves. A group looks like the finest float at Carnival. 'Tricolor' leaves resound with green-and-cream stripes splayed out around a rosy red rosette of tighter foliage wound at the plant's center. The rosette forms a cup to keep the growing point hydrated and to cuddle the flowers. The green inner surface sprouts a deep crown of white and blue flowers, each stiff and pointed like meringue peaks on a freshly baked pie. The plants are terrestrial in nature, as are most in this particular group, and so they are content in well-drained garden beds and pots like the petite earth stars. Wildly striped, solid, or copper blushed, these native Brazilian bromeliads deliver optimistic pools of color and light wherever they are grown.*

Plant Family
Bromeliaceae

Other Common Name
Tricolor bromeliad

Bloom Period and Seasonal Color
Green or green-and-pink striped leaves

Mature Height × Spread
Up to 1 ft. × 1 ft.

Summer Snapdragon
Angelonia angustifolia

It had to happen. Beloved as snapdragons can be, their summer season is limited to the temperate zone. Elsewhere, an equally stunning tower of flowers had to gain the moniker, and this Angelonia is it. A sweet, small plant native to the American tropics, summer snapdragons laugh at heat and humidity in the landscape and in pots. Its tropical pedigree can be seen in its bold, colorful, long-lasting flowers and its love for subtropical and tropical environs. Narrow green leaves are shaped like little fence pickets and layer their rows into cone shapes topped with clusters of flowers. The dime-sized blooms open their throats wide, revealing a colored ruffle all around in solid shades of purple, pink, and white, as well as bicolors that look hand painted. The loose flower spikes look airy, but they are tough enough to support the marauding bands of bees that love them too.

Plant Family
Scrophulariaceae

Other Common Name
Narrowleaf angelonia

Bloom Period and Seasonal Color
Flowers bloom year-round in frost-free areas, spring through fall elsewhere in purple, white, pink, and combinations

Mature Height × Spread
12 in.–18 in. × 12 in.

When, Where, and How to Plant
Select sites for summer snapdragons that have plenty of sun and ready access to water to ensure compact growth and abundant flowers. These plants do not like wet feet. Prepare a soil for beds and containers that is well drained, organic, and fertile. Where native soils are heavy, dense, or very sandy, amend to improve soil structure, fertility, and drainage before planting. Arrange triangles of summer snapdragons for their best effect in mixed borders. Space plants 4 inches apart in beds. Test drainage in potting mixes before planting and amend if needed. For a good show in mixed pots, make room for three summer snapdragons. Take care to plant at the same level the summer snapdragons were growing originally, water well, and mulch after planting.

Growing Tips
Summer snapdragons are easy to grow when they are given a routine. Look at the plants daily and learn what they need. Water regularly, before the plants have any signs of wilt such as pale leaf color or droop. Fertilize regularly too. Use a soluble flower formula or its granular equivalent every other week. Deadhead faded blooms to bring on more flowers.

Care and Propagation
Be on the lookout for mealybugs and aphids. Cut back plants that become leggy and fertilize. Propagate more plants by 4-inch tip cuttings in the fall or grow them from seed.

Companion Planting and Design
Pair summer snapdragons with licorice plants or bat face cupheas for smart contrast and whimsical colors, or underplant the space beneath their good bedmates hyacinth beans or moonflowers.

Try These
'Angel Mist' and 'Angel Face' are two popular cultivars with a wide range of colors and bicolors.

When, Where, and How to Plant

Taro plants thrive in garden soils that are richly organic and uniformly moist. Amend native soils as needed to increase their organic matter with compost, manure, ground bark, or similar material. Prepare an area large enough to plant a tuber several inches deep, or to be able to plant a young elephant ear at the same level it was growing originally. Plant them 18 to 24 inches apart in garden beds or in containers large enough to accommodate individual plants. Except in the warm subtropics and their native range, elephant ears are not reliably hardy. Dig the tubers and dry them to store or pot them up. The plants will probably collapse and begin to go dormant. Overwinter the pots in a spot where they will not freeze.

Growing Tips

Regular water is the key to growing bountiful elephant ears. Install soaker hoses or another reliable irrigation system for elephant ears in garden beds and use it regularly or automate it. Keep plantings mulched to a depth of 2 inches in beds, 1 inch in pots. Fertilize elephant ears with a complete granular formula at planting and, as organic mulch rots, work it into the soil for additional nutrition. Replenish the mulch as needed.

Care and Propagation

Few pests bother these plants, but watch for slugs and snails on young plants in very wet weather. It is propagated by division.

Companion Planting and Design

A row of elephant ears is a natural edging around pools and ponds. Grow them in front of candlebush trees and red Abyssinian bananas, or alongside lobster claw for textural contrasts.

Try These

Another stunning taro, 'Black Velvet', has prominent purple-black markings.

Thank tropical Asia for this most dramatic plant and its quintessential tropical allure. With almost no trunk, taro has a nearly all green or yellow-green leaf, heart-shaped and big as a car door. The veins are prominent on each leaf and seem to bend toward the viewer, no matter the perspective. All plants are far more than 90 percent water, and taro uses its share, enlarging its leaves after thunderstorms and growing even larger with roots in a pond bank. Indeed, care is advised growing taro in water in the subtropics, where it can invade and disrupt wetlands. This plant is the Pacific yam, or dasheen, as it is variously known, where it is specially treated to render it edible and widely used. In the West, it is deservedly celebrated for its potent ornamental presence. Taro tubers are widely available, as are cultivars with darker and multicolored leaves.

Plant Family
Araceae

Other Common Name
Elephant ear

Bloom Period and Seasonal Color
Green leaves all year in the tropics, from the spring through fall elsewhere

Mature Height × Spread
Up to 4 ft. × 3 ft.

Ti Plant
Cordyline fruticosa

Tropical plants have a reputation for being neon-colorful with wild leaves, and ti plant fits the bill perfectly. Tropical plants with colorful leaves are often very diverse and sometimes downright gaudy, and ti plant is both. In its native Southeast Asia and other very warm climates, the trunk shoots up 10 feet with leaves coming off of it like palms. In temperate zones, ti plant is commonly grown as a younger plant, bushy with leaves. They are rowdy bright pink to the deepest burgundy, green and cream in named selections that are streaked, edged, and splashed with color. Ti plant often comes into peoples' lives as a gnarly souvenir of a trip to Hawaii, where its popularity is unsurpassed. The piece of rhizome or log usually roots and grows, at least for awhile. With good care, ti can live up to its other name, good luck plant.

Plant Family
Asparagaceae

Other Common Name
Good luck plant

Bloom Period and Seasonal Color
Prized for multicolored leaves in green, cream, red, and pink stripes

Mature Height × Spread
Up to 10 ft. × 2 ft.; averages 3 ft. × 1 ft.

When, Where, and How to Plant
Choose a sunny site outdoors year-round in the warmest subtropics and tropics. To grow ti plant everywhere else, keep it potted and sink containers into a mulch bed outdoors in the summer. Like its relatives, plants in the *Dracaena* genus, ti plants grow best in soil that is organic and fertile yet very well drained. It is not drought tolerant but cannot tolerate a saturated root zone. Prepare a soil for beds and containers by combining garden soil or potting mix with finely ground bark or sand, compost, and other organic matters. Plant 8 to 12 inches apart in beds and mixed planters or provide an 8-inch pot for individual plants. Select a container slightly wider all around than the plant and that is wider than it is deep.

Growing Tips
Maintain a regular watering schedule that allows ti plants to dry out slightly between waterings. Consistent fertilizing will maintain the strong leaf colors. Use a complete, all-purpose fertilizer four times a year or use a soluble fertilizer mixed in the water six times annually. Keep mulch to a minimum in beds and pots. Repot annually as needed to provide space for its vigorous root system. If leaves lose color, move ti plants into more sunlight, water well, and fertilize.

Care and Propagation
Watch out for mealybugs. Propagate by dividing clumps and rhizomes, or from cut stems called "logs."

Companion Planting and Design
Grow ti plants in madly mixed groups with dwarf powder puffs or make one the centerpiece of a design with wax plant spilling down its front.

Try These
C. australis is more yucca-like, with a strong stem and sword-shaped burgundy leaves.

When, Where, and How to Plant

Add height to any sunny or partly sunny site with consistently damp conditions outdoors by planting umbrella plant. It grows best in garden beds, containers, or beside a pond with fertile, organic, well-drained soil and reliable access to water. Provide ample space for these upright vase shaped plants by spacing them 2 feet apart in beds or in containers 12 inches wide and deep. Line a pond bank or water feature with umbrella plants, and then dig them up with the lilies in the fall. Plant at the same depth the plant was growing originally when transplanting or repotting. Soak deeply after planting and mulch around papyrus with 2 inches of organic material such as ground bark. Group with other water-lovers to simplify your maintenance.

Growing Tips

Unless the plants are growing in water or on a pond bank, use soaker hoses or other well-timed irrigation to ensure that the root zone will be watered consistently. Plants adapt well to container culture and can be held with less water over the winter if they are protected from freezing. Fertilize regularly in the spring, summer, and fall with a complete, all-purpose formula. Keep umbrella plants mulched and as the mulch decomposes, work it into the soil and replace it.

Care and Propagation

Few pests hang around umbrella plants. Propagate them by dividing clumps.

Companion Planting and Design

Keep the beds and pots damp for papyrus and companions such as papaya, lobster claw, reed stem orchids and red gingers.

Try These

'King Tut', also listed as dwarf papyrus, reaches 3 feet as an annual plant and grows well in containers. *C. helferi*, a small cousin, can be grown in aquariums.

Tropical Africa gave birth to the umbrella plant, a grand specimen of utility and legend. It was a source of precious paper in the ancient world and is well known as the grassy rushes along the Nile. But it is the sight of such a bizarre whirligig of a plant that cements its relationship with gardeners internationally. Tall, nearly leafless stems are translucent at first, like big straws with limeade inside. By the time they burst open at their tops, the stems look solid enough to hold up the flower heads proudly. This plant has a tight rosette of flowers centered in a wild crown of long, skinny, spidery, wild bracts. The species belongs at pondside, while selections and cultivars can grow in any garden setting. The named Cyperus species have neater habits and these "uptown girls" are also preferred where the species has become invasive, such as in south Florida.

Plant Family
Cyperaceae

Other Common Name
Papyrus, Egyptian paper reed

Bloom Period and Seasonal Color
Evergreen lime to medium green leaves in the tropics and warmest subtropics

Mature Height × Spread
Up to 5 ft. × 6 ft.

Upright Elephant Ear
Alocasia odora

Upright elephant ear embodies all that is tropical with its optimistic, upright form and striking green leaves as big as couch cushions. It is often the start of a love affair with aroids, a friendly name for the plants in this family. This one has strong stems and brilliant, heart-shaped leaves that unfurl upward in a huge whorl. Upright elephant ears are so breathtaking, it is as if one plant could bring all the green in the world. New leaves are almost lime green and glow proudly until they begin to age and are upstaged by more youngsters. Typical aroid flowers arise in the summer on stout stems. The flowers are subtle, cupped bracts caressing the upright compact flower spike within. The seedpods, packed with shiny red berries, are unexpectedly bold. Count on this native of tropical regions of India, Myanmar, and China to bring a smile to garden visitors.

Plant Family
Araceae

Other Common Name
Night lily

Bloom Period and Seasonal Color
Medium green, heart-shaped leaves average 2 ft. × 1 ft.

Mature Height × Spread
Up to 6 ft. × 4 ft.

When, Where, and How to Plant
Choose firm rhizomes and healthy green plants with no mushy parts. Pot up rhizomes whenever they are available to plant outdoors in warm weather. Plant in part shade, part sun, or dappled sun with ready access to water. This is a focal point plant that deserves center stage. Prepare a soil that is organic, fertile, and well drained in garden beds or containers (although these plants *can* tolerate seasonal flooding). Amend native soils and potting mixes to ensure high organic matter content and good drainage. To plant outdoors, dig a hole twice as wide and one-third deeper than the original container. With rhizomes as big as human arms, upright elephant ear needs room to grow. Space plants 3 feet apart in garden beds or grow individuals in a 5-gallon or larger container. Blanket outdoor plantings with 2 inches of organic mulch.

Growing Tips
Water on a very regular basis is the key to success with this plant. Fill a large saucer under potted upright elephant ears and provide reliable irrigation outdoors to enable largest leaf growth. Fertilize monthly with a soluble complete fertilizer or its equivalent granular product. Remove any flowers when they appear at midseason to stimulate more leaf production. Later in the year, leave flowers on the plant to form seedpods.

Care and Propagation
Few, if any, pests bother these plants. Propagate by dividing their rhizomes.

Companion Planting and Design
Match the *wow* factor of upright elephant ears with dark green split leaf philodendrons, ruffle palms, and white bat flowers. Or make one the centerpiece of a bed of caladiums.

Try These
'Calidora' has very pointed leaves. Look for variegated forms patterned with darker green, yellower, or silver leaf patterns, like 'Okinawa Silver'.

Wax Begonia
Begonia × semperflorens-cultorum

When, Where, and How to Plant
Start seeds for wax begonias indoors six weeks before transplanting to the garden and containers, or purchase small plants when they become available in the spring. Select a warm, partly shady site or, for dark green- and bronze-leaved types, a spot in dappled sun. Use wax begonias *en masse* as low-maintenance edging plants and, in mixed pots, let them fill the space between upright and trailing plants. Prepare a well-drained, richly organic soil by amending garden beds and potting mixes with compost and finely ground bark. Space wax begonias 6 inches apart, plant at the same depth they were growing, and mulch around the plants with 1 inch of organic matter. As the mulch decomposes, work it into the bed or pot and replace it.

Growing Tips
Remove flowers that appear on very young plants to stimulate new growth. Water regularly; let wax begonias dry out only slightly between waterings and use a soluble formula fertilizer monthly. Use an all-purpose product year-round, or substitute a flower formula in the summer. The plants are not drought tolerant and will look pale if they dehydrate. Soaker hoses provide slow, deep watering that encourages deep rooting and stronger top growth. Pinch plants to encourage compact shapes and more flowers.

Care and Propagation
Watch for slugs and snails on tender young plants. Propagate by tip cuttings if you want more plants (you probably will!). See "Propagation" (page 36) for details.

Companion Planting and Design
Tuck wax begonia plants everywhere for instantly tropical tone in garden beds and pots. Good companions include rattlesnake plants, caladiums, justicia, and nun's orchids.

Try These
Popular for decades from the Cocktail series, 'Whisky' has bronze leaves and white flowers.

Diminutive in size compared to many tropical plants, wax leaf begonia nevertheless makes it presence known. Wax begonias may be the most accommodating tropical plant, even though as a hybrid it has no natural habitat to explain its friendly ways. Gardeners can find themselves ankle deep in glowing red, pink, or white flowers in a matter of weeks. Each bloom begins as a tight oval tucked into a leaf cluster, then the bracts unfurl around the true flowers. The plants are compact and look muscular, with fat, round leaves whorled about a thick stem. The leaves are usually a politely glowing green, and they can be bronze or yellow with green variegation. The pièce de résistance in this nearly perfect plant is its amazing self-cleaning habit. As the flowers fade, they dehydrate and drop neatly into the leaves below. No gardener could ask for more.

Plant Family
Begoniaceae

Other Common Name
Fibrous begonia

Bloom Period and Seasonal Color
Shades of pink, red, and white blooms, primarily from the spring through fall

Mature Height × Spread
Up to 18 in. × 18 in.

Wax Plant
Hoya carnosa

Wax plant should be known as the most popular plant nobody knows. To paraphrase Yogi Berra, "Nobody grows wax plant anymore. It's too popular." The famous Yankee catcher was explaining that he no longer went to a restaurant because it was too crowded. For many years, wax plant suffered the same fate. Its vines could be found in homes and arboretums but seldom for sale at nurseries. This carefree plant was considered too common to be commercially viable. Like Yogi's restaurant, wax plant's reputation deterred some of its fans. The succulent leaves of wax plant are oval and dark green, sometimes edged neatly in cream, yellow, or pink. They sprout from surprisingly woody stems that sometimes defy all training efforts but always bend neatly to show off the flower clusters. The flowers are borne in perfect orbs of pink and white stars that seem shockingly fragrant coming from such a rugged, succulent plant.

Plant Family
Apocynaceae

Other Common Name
Wax flower, hoya

Bloom Period and Seasonal Color
Pink and white flowers year-round on mature plants

Mature Height × Spread
Up to 20 ft. vine in the tropics; averages 10 ft. elsewhere

When, Where, and How to Plant
Choose a sunny site for this plant. It grows best in very well-drained soil that is also organic and fertile. It is not drought tolerant but cannot tolerate wet soil for prolonged periods. Where those conditions are impossible to meet, grow it in a container, even in the tropics. Prepare a soil for beds and containers by combining garden soil or potting mix with finely ground bark or sand, compost, and other organic matters. Nestle wax plants down into the soil to make sure they will be stable and provide wires for plants to grow up or space for them to trail. When conditions are warm, it needs high humidity to retain its color. In cooler temperatures, such as indoors, it can tolerate lower humidity.

Growing Tips
Put wax plants on a watering regime that allows the plants to dry out between waterings. Fertilize regularly with a complete soluble formula that includes major and minor elements as well as trace minerals. Use it at full strength in the spring and summer, half-strength in the fall and winter, or use a comparable granular fertilizer four times annually. Repot annually in a slightly larger pot to keep plants growing. If leaves turn pale, move wax plants to a sunnier location.

Care and Propagation
You will need to watch for aphids and mealybugs. They can be propagated by cuttings.

Companion Planting and Design
Use wax plants as a centerpiece for sunny beds, in groups of hanging baskets, or spilling out of a window box. Plant them among kanga paw, ti plants, and Cape daisies.

Try These
'Krinkle Kurl' has twisted ropes of leaves. *H. serpens* is a tiny plant with white flowers that turn pink.

White Bat Flower
Tacca integrifolia

When, Where, and How to Plant
White bat flowers grow well outdoors in the subtropics and tropics in consistently damp soil and shady conditions, or they can be grown in a container all year regardless of the climate. The soil should be richly organic and fertile, yet be able to drain well enough to prevent a saturated root zone. Amend garden soils and potting mixes by adding organic matters including peat moss for added acidity. Space white bat flower plants 2 feet apart in beds and mixed planters to ensure good air circulation around the plants. Plant at the same level it was growing originally or slightly higher. Grow one plant in a standard 10-inch clay pot and provide a saucer to water it from the bottom. Mulch plantings well with organic material such as ground bark.

Growing Tips
Water white bat flowers very regularly to keep their soil consistently moist, but avoid overwatering in the winter. Fertilize four to six times each year with a balanced, soluble formula added to the water. Indoors, these plants enjoy light, temperature, and humidity conditions in the human comfort zone. Expect the flowers to last for weeks, but clip them off when they are done. If a leaf dries up occasionally, remove it. If several leaves die, suspect root rot.

Care and Propagation
There are few pests to bother it. White bat flowers are propagated by divisions.

Companion Planting and Design
Contrast white bat flowers with colorful leaves of prayer plants and nerve plants. Grow them for drama in a group with striped blushing bromeliads and ferns, from Boston to bird's nest ferns.

Try These
T. caulteri is the black bat flower, an eerie reverse of the white bat flower with darker aromas.

Native to India and Southeast Asia, white bat flower demonstrates the lengths a plant will go to accommodate its pollinators and so ensure future generations. The leaves are commonplace, held on short green stems and look like gray-green, curving cousins of the cast iron plant. Such nondescript foliage hosts what may be the most otherworldly flower in the tropical garden. Each blossom opens atop a stiff stem in a cluster of two dozen purplish brown, tightly wrapped flowers that form the bat's face with two white bracts held stiffly above the flowers to create the bat's wings. That should be enough to capture your attention, but these flowers are bearded with dozens of long, elegant tendrils that cascade down to leaf level. Although white bat flower has no peers in the bizarre beauty department, it is no plant snob. The plant thrives and blooms indoors alongside the ever-popular peace lily and reliable butterfly orchid.

Plant Family
Diascoreaceae

Other Common Name
None

Bloom Period and Seasonal Color
White and purple flowers from late spring through fall and intermittently in winter

Mature Height × Spread
Up to 4 ft. × 2 ft.

Woody Tropicals

อินทนิน

Angel Trumpet
Brugmansia suaveolens

Make room for loud, proud angel trumpets to parade into the garden like a marching band. They are vibrant, bold, and easy to grow. Tall, sturdy canes rise from ground level, naked and gnarly. The canes need strength to support the umbrella of huge leaves that erupts from their tops. Each leaf can be as big as a boat paddle, coarse to look at, and rough to the touch. The leaf canopy sustains and displays the flowers, long trumpets with wide mouths that hang straight down from its branches. Classic angel trumpets are single white or gold, while hybrids and selections deliver a rainbow of colors plus ruffles, fringes, and fully double flowers. The plants quickly transform any sunny space into a flowery, tropical oasis. With an innate ability to go dormant in its native subtropics, angel trumpets readily adapt to overwintering indoors in the temperate zones.

Plant Family
Solanaceae

Other Common Name
Angel star

Bloom Period and Seasonal Color
Summer and fall flowers in shades of yellow, white, salmon, purple, and pink

Mature Height × Spread
Up to 8 ft. × 5 ft. on average; larger and smaller types are not uncommon

When, Where, and How to Plant
Warm sun, well-drained soil, and regular watering are the keys to growing angel trumpets. Choose the sunniest site available to enable the earliest flowering. Prepare a soil for garden beds or pots that is fertile, organic, and well drained. Amend with organic matter and sand if necessary to ensure great drainage and to improve its water-holding capacity. Space angel trumpets 2 to 4 feet apart in beds or plant in 10- to 15-inch pots, depending on the variety. Add ½ cup composted manure per plant in the hole at planting. Water deeply and slowly to encourage deep roots to support these tall plants. Skip any mulching until the soil warms to let sunlight warm the young plants from the ground up in temperate climes. Establish a regular irrigation pattern and use it. *All parts of this plant are poisonous. Do not plant where it can become an attractive nuisance.*

Growing Tips
Water regularly and do not allow the plants to wilt between irrigations. In hot weather, mulch around the base of angel trumpets or provide a living mulch with licorice plants or other low growers. Fertilize angel trumpets regularly with a granular or soluble all-purpose fertilizer. Begin when new growth emerges in late winter and apply as often as the product label advises. Deadhead flowers as they shrivel to prevent fungus development.

Care and Propagation
Watch for chewing insects and spider mites. Propagate from semi-hardwood in the summer or by rooting the tips of canes in water after the leaves shed in the fall.

Companion Planting and Design
Let angel trumpets tower over chenille plants, cat's whiskers, and bat face cupheas. Plant a tall group with frangipani, joy perfume trees, and bird of paradise.

Try These
Perfect for pots, 'Charles Grimaldi' opens yellow and turns warm orange as its flowers age.

Arabian Jasmine
Jasminum sambac

When, Where, and How to Plant

Arabian jasmine plants grow best in beds and pots with fertile, organic, well-drained soil. They are not at all drought tolerant and need water consistently, but they cannot tolerate a flooded root zone either. Amend existing garden soils and potting mixes with compost, ground barks, and other aged organic matters such as manure to meet these needs. Space the plants 1½ feet apart in beds, or plant two in a container 12 inches wide and deep. When planting or repotting, do not bury the stems. Be sure the jasmine plants will be growing at the same level as they were before. Cut back new plants to stimulate branching and begin shaping their ultimate size and form. Mulch all plantings with 2 inches of organic material.

Growing Tips

Plenty of water applied deeply and regularly will keep these plants in good health. If consistently damp soil cannot be maintained, consider reservoir pots or automated irrigation. Fertilize Arabian jasmines regularly while they're young, then less frequently. Use soluble fertilizer mixed in water in the early years, and then continue with a complete granular formula three or four times each year. As organic mulch rots naturally, work it into the soil for additional nutrition. Prune regularly after it blooms to keep the plants compact.

Care and Propagation

Arabian jasmines suffer from few pests. Propagate them by tip cuttings.

Companion Planting and Design

Arabian jasmines can fill the space around ruffle palms for green contrast and floral scent. Grow them with Philippine orchids and fragrant heliotrope, or with sanchezias for a stunning purple and gold theme.

Try These

'Grand Duke of Tuscany', also known as 'Flore Pleno', has gardenia-like, double flowers.

Probably native to India, the ironically named Arabian jasmine been cultivated for so long that no one knows exactly where it grew first! The plants bloom perpetually in the tropics and will be returning perennials in the warm subtropics and even a bit farther north in sheltered locales. In the latter situations, its flowers are delayed, which explains their popularity as a container plant. This plant is one tough character compared to other jasmines, stuck between a vine and a shrub with gangly stiff stems. Such an unsophisticated air could limit its appeal, but Arabian jasmine cleans up nicely with regular pruning and becomes downright glamorous when it blooms. Waxy white flowers with overlapping petals exude a thick scent that continues to waft through the garden even as the blossoms age to a sweet pink color. It is no wonder this plant is the national flower of the Philippines, where it is known as sampaguita.

Plant Family
Oleaceae

Other Common Name
Sampaguita

Bloom Period and Seasonal Color
White flowers year-round

Mature Height × Spread
Up to 10 ft. × 4 ft.; averages 6 ft. × 3 ft.

Bamboo Palm
Chamaedorea seifrizii

The rainforest floors in Belize and Honduras gave birth to bamboo palm, an uptown girl of a plant who knows how to strut her stuff. This palm has risen to the pinnacles of plant prestige, chosen for many public and private atriums and sunrooms. A clump of gray-green stems, reedy and jointed, sports thin stems and long, pointed arrow-shaped, evenly pinnate leaves. They are stunning and graceful, spaced as if to stir the air on a sultry day. Bamboo palm might be the easiest tropical plant to grow and is among the most rewarding, once its needs are understood and met. Outside the tropics, this plant and its relatives can often be seen in dark corners, covered with dust and yet still standing, outlasting their owners' efforts to overlook them permanently. With water when the soil is dry and a shaft of light, bamboo palm readily grows into an elegant, long-lived plant.

Plant Family
Arecaceae

Other Common Name
Reed palm

Bloom Period and Seasonal Color
Jointed canes (like bamboo) and evergreen darker green leaflets

Mature Height × Spread
Up to 10 ft. × 3 ft.; averages 5 ft. × 2 ft.

When, Where, and How to Plant
Warm spaces in shade, bright light, and dappled shade readily accommodate bamboo palm. Out of its native range, to meet its needs for very well-drained soil and undisturbed roots, bamboo palm is best grown in a container. Sink one into a mulch bed for summer's quickest tropical tone. Avoid areas where salt spray or constant wind will damage the leaves. Prepare a potting mix that drains very well yet is both fertile and richly organic. Add ground bark and compost as needed to achieve these goals. Mulch palms with no more than 1 inch of organic material. Bamboo palms readily move indoors for the winter without leaf drop or drooping.

Growing Tips
Inside or outdoors, water this palm well and repeat when the top of the soil feels dry under its light mulch covering, but no sooner. Fertilize monthly in the spring and summer with a soluble general-purpose formula mixed at half strength, less often in the fall and winter. Lift sunken pots one month before your area's first frost. Inspect them for insects, groom the plants and, if roots have grown out of the pot, repot before transitioning to indoors.

Care and Propagation
Mist the undersides of leaves occasionally to deter spider mites. Propagate by divisions and seed. See "Propagation" (page 36) for details.

Companion Planting and Design
Bamboo palms tower in the moderate conditions and look good surrounded by bush lilies and multicolored crotons or snake plants and queen of the night for diverse colors and textures.

Try These
C. elegans, parlor palm, is a smaller cousin, as is *C. metallica* which has a deep blue-green, metallic shimmer.

Bottlebrush
Callistemon viminalis

When, Where, and How to Plant
Select a site in full sun with access to water. A sheltered location, such as in front of a warm wall, is ideal for locations in the subtropics and container grown plants. This bottlebrush can adapt to a variety of soil conditions. Amend heavy clay soils and rich potting mixes with equal parts ground bark to improve drainage while retaining water-holding capacity. Water regularly when growing young bottlebrush; mature plants can take some flooding. For best performance, take care not to overwater mature plants and those spending the winter indoors or in holding areas. Space bottlebrush plants 3 feet apart in beds or large planters or provide a container at least 15 inches wide and deep for one plant. Mulch new plantings.

Growing Tips
Keep plants watered in the spring so the flowers open fully. After that, allow the plants to dry out slightly between irrigations. Fertilize in the spring and summer with a granular all-purpose formula or a flowering formula if needed to stimulate blooms. Prune as needed to establish and maintain a weeping form, including removing the lower branches and shortening those that drag the ground. Clip off spent flowers to encourage new leaves and longer branches.

Care and Propagation
Watch out for spider mites and sawfly larvae. Propagate by semi-hardwood cuttings.

Companion Planting and Design
Use a bottlebrush as a centerpiece on an island surrounded by musical note plants and gazanias. Good companions for height in container gardens are miniature date palms, mallow, and hibiscus.

Try These
'Little John' is a dwarf form, while 'Hannah Ray' has graceful weeping form.

This bottlebrush is native to New South Wales, Australia, and is adapted to the warmest parts of the United States. Some relatives are hardier, but none is as dramatic in the tropical garden. Leaves begin bronzy green and develop darker shades as the summer approaches. Lovely, long, narrow, and pointed, they line up along stiffly arched branches. They seem to punch the air, pushing aside anything in their way as they prepare to display awe-inspiring flowers. Branch tips soon turn to fat, scarlet red blooms the size of juice cans. The blossoms do look like brushes, as if they were drawn by a cartoonist. Scores of brilliant stamens cover the outer surface of the flowers that hang like extravagant earrings. Bottlebrush flowers are dense and seem heavy until the slightest breeze lifts them up and down flirtatiously. Their dazzling dance is irresistible to hummingbirds as well as humans.

Plant Family
Myrtaceae

Other Common Name
Red bottlebrush

Bloom Period and Seasonal Color
Red flowers from the spring into summer, year-round in frost-free areas

Mature Height × Spread
Up to 20 ft. x15 ft. in tropics; up to 8 ft. × 5 ft. elsewhere

Candle Bush Tree
Senna alata

Unless an old timer could be persuaded to give up some seeds, candle bush tree used to be unavailable to gardeners outside the tropics. A large tree in its hot, wet, native homes in the American tropics, it adapts well to life as an annual plant elsewhere and may become perennial in the subtropics. The leaves look curiously neat, fanning out flatly opposite one another to create long leaflets. The flower clusters look like golden, cottony balls stacked straight up, each 8 inches high. The plants branch to make room for these glowing blooms to create a breathtaking garden candelabrum. Cottage gardeners in temperate zones still prize the plants, their flowers, and the big, precious seedpods that ensure next year's planting. Fortunately for everyone else, both seeds and plants have enough devotees these days to keep candle bush tree growing everywhere.

Plant Family
Fabaceae/Leguminosae

Other Common Name
Candlestick tree, Christmas candle

Bloom Period and Seasonal Color
Rich golden flowers in late summer and fall

Mature Height × Spread
Up to 12 ft. × 4 ft.; averages 6 ft. × 3 ft. as an annual plant and in pots

When, Where, and How to Plant
Select sites that have plenty of sun and ready access to water to enable steady growth and abundant flowers. Prepare a soil for beds and containers that is well drained, organic, and fertile. Where native soils are heavy, dense, or very sandy, amend them to improve their soil structure, fertility, and drainage before planting. Start seeds indoors in large peat cups six weeks before the last frost is predicted, or purchase small plants. Plant one tree in a container that is 8 to 12 inches across and equally deep for stability as it grows. Space young trees 2 feet apart in garden beds. Mulch with 1 inch of organic material after the soil is very warm and growth is well under way.

Growing Tips
Young plants are easy to grow with a simple routine of water, fertilizer, and more water. Plants will mature in frost-free areas and become quite drought tolerant. They cannot stand in water for long but need plenty of it to produce enough leaves in the first summer to flower grandly. Soaker hoses or drip irrigation can make this task much easier. Stake young plants if needed to direct their upright growth and pinch back the tips to encourage branching. Fertilize with a soluble or granular all-purpose formula as often as the product directs.

Care and Propagation
Keep your eyes open for mealybugs and spittle bugs. Propagate by seeds or cuttings in late winter.

Companion Planting and Design
Make room for a candle bush tree between two sago palms in pots on a sunny porch. Before it blooms, ring the tree with colorful plants such as coleus, scarlet sages, and fan flowers.

Try These
Tall and woody, golden shower tree (*C. fistula*) dazzles in late spring.

Dragon Tree
Dracaena draco

When, Where, and How to Plant

A site in full or dappled sun outdoors and a sunny room indoors can provide ample light for dragon tree. Outside the dry tropics, it is best grown in a container for most of the year and moved indoors in freezing weather. Provide an organic, fertile, very well-drained soil by amending garden soil and potting mixes with compost or ground bark. Plant one small dragon tree in a standard 10-inch clay pot, moving up to larger containers and fresh soil annually in late winter as needed. Mulch plantings lightly, with no more than 1 inch of organic matter. Over time, dragon tree will become a large, heavy plant. Use rolling plant saucers or a hand truck to move the plant without back injury.

Growing Tips

Overwatering can cause dragon tree to rot. Allow pots or plantings to dry out between waterings. Dig up flooded plants and repot immediately in fresh, well-drained soil. Fertilize monthly with a soluble, general-purpose formula as directed on the product label in the spring and summer, then at half strength in the fall and winter. Groom dragon tree to remove old leaves and keep the leaf clusters clean and free of debris that falls in to their depths.

Care and Propagation

Be on guard against aphids and scale insects. To propagate more plants, take cuttings. See "Propagation" (page 36) for details.

Companion Planting and Design

Grow dragon trees with contrasting reeds and leaves such as bamboo palms, crotons, and laurel figs. Combine them with equally strong leaf shapes of snake plants and jade plants.

Try These

D. marginata 'Tricolor' has a softer form, narrow leaves striped pink, green, and cream.

Native to the Canary Islands, dragon tree stands as erect and steadfast as a palace guard overlooking all it surveys. Primly upright, the trunk is thick and gray, sometimes gnarly, chunky, and slick to the touch. Like its relatives the agaves, this dracaena has sword-shaped leaves held close on very short stems; unlike them, all the leaves fan out from the top on this one, lifted on a handsome rack of branches atop the trunk. Among the most dramatic of its family, dragon tree is also among the largest and towers grandly in dry tropical landscapes. More good-natured than it looks, dragon tree makes a stunning, low-maintenance addition to the indoor garden. Mature plants can bloom with fine yellow flower clusters followed by striking red-orange seedpods. The tree exudes a red sap when cut. According to European myths, this sap is dragon's blood, to be feared and extolled.

Plant Family
Asparagaceae

Other Common Name
Dragon's blood dracaena

Bloom Period and Seasonal Color
Medium green leaves are evergreen in warm environments

Mature Height × Spread
Up to 30 ft. × 10 ft.; averages 6 ft. × 2 ft.

Dwarf Powder Puff
Calliandra haematocephala 'Nana'

From Honduras and southern Mexico comes a shrubby plant that packs a powerful punch of fast growth and brilliant colors. Dwarf powder puff plant is draped in layers of shimmering leaves that emerge reddish bronze and open to bright green. This versatile shrub grows into a dense bush in full sun or develops a looser form in partly shady places. Its flowers are abundant and truly mysterious. Red buds pop out that look more like berries or seedpods than future flowers. From each raspberry-shaped pod comes a fat, red, fan-shaped pompon about 3 inches wide. Not only does the flower look like a powder puff, it feels oddly soft to the touch as well. The flower show goes on all year in frost-free climates, including sunrooms and greenhouses, and the plants may become perennial in the subtropics.

Plant Family
Leguminoseae

Other Common Name
Flame plant

Bloom Period and Seasonal Color
Flowers from the summer to fall

Mature Height × Spread
Up to 8 ft. × 5 ft.; averages 5 ft. × 3 ft.

When, Where, and How to Plant
Choose a site in sun, part sun, or part shade and plan to water regularly in warm weather. Prepare a soil for dwarf powder puff plant that is fertile and very well drained. Add ground bark and sand to garden soils and potting mixes to improve these qualities. Combine one part organic matter and one part sand with three parts existing soil. Space plants 2 to 3 feet apart in beds and patio planters, or plant one in a 5-gallon nursery pot. Water new plants well and mulch with organic material. Dig up garden plants in the fall if necessary and overwinter indoors or in an outdoor shed. As long as the rootball does not freeze, the plants will sprout anew when the weather warms up again.

Growing Tips
Dwarf powder puffs are not small, nor are they drought tolerant. Water regularly in warm weather but do not let the plants become flooded. Maintain 1 inch of organic mulch around the base of the plants. When the mulch rots, work it into the planting and replace it. Fertilize in the spring and summer with a granular product that has both major and minor elements included. To encourage hardiness in the subtropics, mulch heavily when temperatures fall below 45 degrees F.

Care and Propagation
Be on the lookout for whiteflies and scale insects. If you like, propagate it by cuttings and seeds.

Companion Planting and Design
Design for contrasting textures by pairing dwarf powder puffs with devil's backbone, ti plants, or musical note plants. Grow it in a big pot surrounded by a skirt of African daisies or gerberas.

Try These
Named selections include many shades of red and pink flowers.

Dwarf Ylang Ylang
Cananga odorata

When, Where, and How to Plant

Dwarf ylang ylang can live its entire life happily in containers or be planted in garden beds for the summer and returned to pots in the fall. Choose a site that gets sun most of the day, or most of the day, indoors all year and outside in warm weather, always with ready access to water. Prepare a soil that is richly organic, fertile, and well drained, even sandy. If necessary, amend garden soils and potting mixes to ensure these qualities. Plant one young tree in a container 12 to 14 inches across and about as deep, or space plants 2 to 3 feet apart in beds. Keep 1 to 2 inches of organic mulch around the plants at all times. Support young trees with thin stakes if needed for the first few months to prevent them from bending.

Growing Tips

Provide water very regularly to dwarf ylang ylang trees throughout the year except when sunlight is very limited. Fertilize plants in the spring, summer, and fall with an all-purpose formula that includes major and minor elements as well as trace minerals. Remove faded flowers unless the seedpods are desired. Gardeners who enjoy pruning appreciate that it blooms on new wood, and regular light pruning helps to keep it blooming sporadically all year in the right conditions.

Care and Propagation

Dwarf ylang ylang trees suffer from few pests. Propagate by cuttings taken from semi-hardwood if you want more plants. See "Propagation" (page 36) for details.

Companion Planting and Design

Paint a scene in shades of yellow by pairing it with candelabra trees, yellow cestrums, and yellow shrimp plants. Let dwarf ylang ylang deliver strong contrast to African gardenias and bird of paradise plants.

Try These

'Fruitosa' is a stalwart favorite that is ideal for container growing.

Dwarf ylang ylang trees cannot be ignored when their hypnotic fragrance drifts across the garden at night. It lives up to its exotic name, adapted from the Tagalog for "flower of flowers" with unforgettable ribbonlike bloom clusters. The ylang ylang tree is native from Myanmar and Malasia, through the Philippine Islands to northern Australia where it is a huge tree. Its large leaves are ribbed rather like avocado leaves and curve slightly down, all the better for showing off nodding blooms. The flowers are curly streamers that look as mysterious as their aroma; they have long been used in leis and are said to an aphrodisiac for women. Perhaps that is why famed fashion designer Coco Chanel chose fragrance No. 5 from a selection presented for her consideration by Russian perfumer Earnest Bo in 1921. Ever since, ylang ylang has been a prized essential oil in perfumery and aromatherapy.

Plant Family
Annonaceae

Other Common Name
Chanel plant

Bloom Period and Seasonal Color
Golden yellow flowers off and on all year in sunny warm conditions

Mature Height × Spread
Up to 8 ft. × 4 ft.

Fairy Petticoat
Elaeocarpus grandiflorus

When fairy petticoats bloom, and then bloom some more, it is possible to believe in Tinker Bell. When their floral skirts continue their dance for months and months, the possibilities seem endless. The plant is reminiscent of oleander, shrubby with gray stems and covered in strappy green leaves that have narrow points on each end. When the flowers fling their fat pink buds into midair, they can cover the stems almost entirely, opening into a huge skirt of white- and pink-fringed bells. Because the plants bloom for months, the buds, flowers, and dusty blue berries can be present on the plant at the same time in a truly spectacular display of its tropical nature. As exotic looking as its origins in tropical Asia and Australia, fairy petticoat nonetheless adapts well to indoor microclimates that allow humans a close-up view of the earliest flower shows.

Plant Family
Elaeocarpaceae

Other Common Name
Lily of the valley tree

Bloom Period and Seasonal Color
White- and pink-trimmed flowers winter to summer, sporadically after that

Mature Height × Spread
Up to 15 ft. × 4 ft.; averages 5 ft. × 2 ft. in containers

When, Where, and How to Plant
Like shower orchids, fairy petticoat plants bloom beginning in late winter and thus are best grown in containers outside the tropics. Outdoors, keep pots in warm sun away from salt spray and persistent breezes to extend the life of the flowers even more. When temperatures fall below 55 degrees F, move the pots indoors to a room with moderate temperatures and bright light. Fairy petticoats grow best in a soil that is very well drained, organic, and fertile. Potting mixes alone may be too dense for this plant and can benefit from amendments of ground bark to improve drainage. Provide a standard 8- to 12-inch clay pot for each plant. Do not mulch fairy petticoat plants except in very dry environments. Support young plants as needed for stability.

Growing Tips
Water regularly, but allow the plants to dry out between waterings. The plants are not drought tolerant and cannot withstand saturated soils. Use a soluble fertilizer mixed at half strength monthly from late fall through summer. Choose an all-purpose product or one made specifically for flowering plants. Prune to shape after the heavy flowering is finished. Prune each year to avoid heavy pruning at any one time and subsequent poor flowering.

Care and Propagation
It has few insects to bother it, but watch for aphids on new growth. Fairy petticoats can be propagated from cuttings.

Companion Planting and Design
Flaming glorybowers, reed stem orchids, Philippine orchids, calico plants, and butterfly flowers all work well with fairy petticoats.

Try These
Shower orchid (*Congea*) also blooms in late winter and can be grown much like fairy petticoat.

False Aralia
Schefflera elegantissima

When, Where, and How to Plant
Grow false aralia in shady places or those where there is mostly shade and some brighter light. Avoid direct sun, indoors or outside. This plant can spend its entire life in a container or planted in beds outdoors in the tropics and in atriums elsewhere. Container plants benefit from summers sunk into mulch beds in shade or dappled shade. Provide an organic, very well-drained soil by amending garden soil and potting mixes with compost or ground bark. Plant one false aralia in a standard 10-inch pot or space several 8 to 12 inches apart in rows and beds. Mulch plantings lightly in beds and not at all in pots. If garden soils are likely to stay wet for months at a time, grow in pots.

Growing Tips
Overwatering can be fatal to false aralias. Do not allow pots or plantings to stand in water. Allow pots or plantings to dry out before watering again. Fertilize regularly with a soluble, complete formula. Use it at full strength while the plants are actively growing, then less often in the winter and once plants mature. False aralia will grow more steadily if it is repotted annually into a slightly larger pot. Clip off the canes if they dry out and water slightly more often.

Care and Propagation
Watch for aphids and spider mites. Propagate by division.

Companion Planting and Design
Let a false aralia be the focal point among bush lilies, earth stars, and night-blooming cereus, whether they are in bloom or not. Go for a green scene with jade plants and snake plants.

Try These
A good companion and a close relative, dwarf schefflera tolerates human foibles almost as well as false aralia.

To make it into the top ranks of popular tropical plants, one must be reliable to a fault, express a memorable style, and keep their passports up-to-date—that is, be able to travel without disaster to markets far from home. False aralia does it all with dark, menacing grace that has spread it from the South Pacific to gardens on every continent. Its unforgettable juvenile leaves are such a dark green that they are nearly inky black, with lighter midribs and brownish red undersides. Each leaf is lobed wildly, cut like hands and fingers aflutter with embarrassment. They stick out in every direction all along the length of their elegant canes in a perpendicular geometry that cries out for attention. False aralia tolerates a range of conditions yet proudly waves the tropical banner of bizarre leaf color, memorable shape, and a presence that will not—and should not—be ignored.

Plant Family
Araliaceae

Other Common Name
Threadleaf aralia

Bloom Period and Seasonal Color
Evergreen leaves

Mature Height × Spread
Up 25 ft. tall and wide; averages 5–12 ft. × 3–5 ft. outside the tropics

Firecracker Plant

Russelia equisetiformis

Native to Mexico, firecracker plant is a favorite among gardeners looking for alternative textures in the tropical world of huge leaves and fat flowers. Somewhere between a shrub and a reedy perennial, this plant marches to its own drumbeat, a raucous cadence of vigorous, wiry lines. Firecracker plants can be comfortable in garden beds as far north as the subtropics but are equally at home in hanging baskets. The bending stems look nearly naked, the leaves simplified to mere scales packed tight around them. Bloom clusters are 1-inch-long, spicy red tubes—firecrackers—that pop open with yellow lips. Their distinctly different appearance, almost leafless yet bursting with flowers, makes it seem as if the plant had to shed its foliage to gain the bountiful blooms. But no, firecracker plant brings a rare angular form to the garden's design, starkly beautiful and boisterous as it summons the butterflies that hover around it constantly.

Plant Family
Plantaginaceae

Other Common Name
Coral plant

Bloom Period and Seasonal Color
Coral flowers year-round

Mature Height × Spread
Up to 6 ft. × 2 ft. in frost-free areas

When, Where, and How to Plant
Select a site in sun or part sun with a reliable water source nearby to grow young firecracker plants. Once established, they are quite drought tolerant. Provide a soil for garden beds or containers that is organic, fertile, and well drained. Space plants 1 foot apart in garden beds and large planters, or plant three small plants in a 10-inch hanging basket. These plants flourish in garden soil in warm weather everywhere, are hardy into the subtropics, and are not difficult to maintain over winter as potted plants in the temperate zone. Lay a blanket of organic mulch in beds and pots that is 1 to 2 inches deep.

Growing Tips
Water and fertilize firecracker plants regularly and let them dry out just slightly between irrigations. Use a general-purpose garden fertilizer as often as directed on the container from the spring through fall and at half strength in the winter or if no additional size is desired. Prune stems to the ground after flowering to promote new growth. Maintain organic mulch around the plants and work it into the soil as it rots naturally.

Care and Propagation
There are few pests that bother firecracker plants. Propagate them from tip cuttings anytime new growth is present. See "Propagation" (page 36) for details.

Companion Planting and Design
Give firecracker plants the equally dramatic bedmates of Brazilian red cloak, copperleafs, and yellow bells. Plant them between angel trumpets and golden dewdrop for contrasting lines in the design.

Try These
R. sarmentosa, red rocket, has red tubular flowers in neat whorls.

Flame of the Woods

Ixora coccinea

When, Where, and How to Plant

Choose a site in sun, part sun, or even part shade for this versatile plant, although full sun should be reserved for temperate zone gardens. The plants thrive in a soil that is organic, fertile, and well drained. Ixora can tolerate some salt spray but it will not grow in alkaline conditions. Combine garden soil or potting mix with compost and other organic matters to meet the plant's needs for an acid soil. Space plants 18 inches apart in beds or provide a standard 10-inch clay pot for one flame of the woods. Water new plantings well and use root stimulator fertilizer to prevent transplant shock. Mulch plantings in beds or pots lightly, if at all, with no more than 1 inch of organic matter.

Growing Tips

Give flame of the woods routine maintenance that allows the plants to dry out between waterings. Water well and then wait to water again until the soil looks and feels dry on its surface. Use a granular, complete, all-purpose fertilizer around the base of the plants four times each year or use a soluble formula every other month. Repot container grown plants annually as needed to accommodate growing plants.

Care and Propagation

Watch out for aphids and mealybugs, and sometimes for scale insects. Propagate by tip cuttings of semi-hardwood if you want more plants.

Companion Planting and Design

A collection of colorful ixora plants makes a fine complement to hibiscus and offers contrast to evergreens such as foxtail agaves. Surround flame of the woods with gerber or Cape daisies.

Try These

'Nora Grant' is a dwarf and 'Super King' is a mammoth plant with very large flower clusters.

This plant is native to India and Sri Lanka. Shrubby, with plenty of branches to create a solid form in the landscape, flame of the woods is a prizefighter of a plant. Stocky and compact, leaves arranged in neatly erect opposite pairs, there is not an ounce of foolishness in this plant's foliage, which is deep green and leathery. Flame of the woods has little competition for dynamic shape in many settings. Seen as a focal point in a garden bed, its round flower heads sing out for attention, waxy blossoms gathered into natural ice-cream scoops of color. With clear hues of orange, pink, or yellow taking center stage with jade green new leaves nearby, even one plant is a standout. Grown in a row and neatly clipped in the tropics and warm subtropics, flame of the woods can be a fine alternative to gardenias and other acid-loving hedge plants.

Plant Family
Rubiaceae

Other Common Name
Ixora

Bloom Period and Seasonal Color
Orange, pink, or yellow flowers in summer

Mature Height × Spread
Up to 8 ft. × 6 ft.; usually pruned to 2 ft. × 2 ft.

Foxtail Agave
Agave attenuate

For the bold shape and cool color of an agave without the usual spines, meet foxtail agave. Leaf colors range from light green to medium gray, and they are narrow at their base, widening broadly before tapering to a point. With old age come the flowers, powering up from the rosettes of leaves like green gushers with pointed yellow-green peaks. Native to central western Mexico, foxtail agave is rare in the wild but has been widely adopted by gardeners. A plant's popularity can be seen in the numbers of common names it garners and this one has three more: swan's neck, in celebration of its sensually curved stems; lion's tail, for its tall flower spike; and less poetically, spineless agave. Do not doubt this plant's reliability, however. It has the "spine" to live for years before climaxing its life with a showy flower spike, and by then it has suckered into a neat clump of future plants.

Plant Family
Asparagaceae

Other Common Name
Swan's neck, lion's tail, spineless agave

Bloom Period and Seasonal Color
Flower spikes are covered in green-yellow flowers

Mature Height × Spread
Up to 5 ft. × 5 ft. with flower spike up to 14 ft.; averages 3 ft. x. 3 ft. in pots with 5 ft. spike

When, Where, and How to Plant
Choose a sunny site for this outstanding container plant, which can grow outdoors year-round in the warmest subtropics as well as in its native tropics. Unlike most of its relatives, foxtail agave grows best in soil that is organic and fertile yet well drained. It is not drought tolerant. Prepare a soil in beds or containers that combine garden soil or potting mix with finely ground bark or sand, compost, and other organic matters. Plant at the same level it was growing before or slightly higher, with ample space around its base to accommodate its curving stem habit. Select a container that is 3 to 4 inches wider all around than the plant, and preferably wider than it is tall for increased stability.

Growing Tips
Keep agaves on a regime that allows the plants to dry out between waterings. Apply a granular, complete, all-purpose fertilizer around the base of the plants four times each year or use a soluble formula every other month. Foxtail agave stems bend naturally with age and growth. Keep mulch to a minimum in beds. Repot annually as needed to provide space for the growing trunk. If leaves turn pale, move foxtail to a location with more sunlight.

Care and Propagation
Few, if any, pests are around to bug foxtail agaves. To propagate, remove suckers from around the base of the plant or grow from seed.

Companion Planting and Design
Grow a pair of foxtail agaves at opposite ends of a group of plants to anchor a scene with red bottlebrush in back and orange kanga paw and white African daisies in between.

Try These
A variegated form of foxtail agave has creamy white stripes on its leaves.

When, Where, and How to Plant

Choose a site where sunny or partly sunny conditions prevail and that has a reliable water source nearby to grow frangipani. Add sand, compost, and other organic matters to existing garden soils and potting mixes to achieve these conditions. Prepare a soil in garden beds or containers that is richly organic and fertile with very good drainage. Space plants 1 foot apart in garden beds and mixed planters, or plant one frangipani in a standard 10- to 12-inch pot. These plants have shallow roots and can be grown outdoors in warm weather and maintained over winter as potted plants in the temperate zone. Mulch frangipani with 1 inch of organic material such as ground bark or pine straw.

Growing Tips

Water frangipani regularly and let the soil dry out only slightly between irrigations. Use a general-purpose garden fertilizer, either a granular or soluble formula, as directed from the spring through fall, and less often in the winter. Prune container plants lightly before bringing them indoors for the winter. Maintain organic mulch around the plants and work it into the soil as it rots naturally. Occasional deep watering will encourage deep rooting and increase stability in tree form plants.

Care and Propagation

Be on guard against mealybugs. If you want more plants, propagate from cuttings taken from semi-hardwood. See "Propagation" (page 36) for details.

Companion Planting and Design

Pair frangipani and hibiscus at poolside with cat's whiskers and fan flowers at their feet. Grow some amid caricature plants and dwarf cone gingers. Contrast its flowers with bamboo orchids and rose cactus.

Try These

'Caroline B' is rosy pink and white, and 'Marion B' has yellow centers and rose-blushed white petals.

Scores of hybrids and selections of frangipani make collecting these plants a never-ending rainbow of shades of pink, rose, yellow, white, and purple. They are hardy only in the warmest tropical environs but readily adapt to container growing. More than 200 named varieties exist, all native to the American tropics, and are grown worldwide for their wide range of fragrances. Various frangipani flowers are said to have aromas like nearly everything, from coconut to citrus, rose, and apricot, and many are staples of Hawaiian lei-making. The small trees have gnarled trunks and broad canopies of big, rather succulent leaves that are deeply veined and almost slick-looking. The flowers are loosely clustered bunches of long tubes that flare open to reveal colorful, waxy corollas that can be solid, bicolored, or blushed with tinges of pink and yellow.

Plant Family
Apocynaceae

Other Common Name
Plumeria

Bloom Period and Seasonal Color
White, pink, yellow, or red flowers sporadically all year

Mature Height × Spread
Up to 20 ft. × 10 ft.; averages 8 ft. × 2 ft.

Golden Dewdrop
Duranta erecta

Sprawled across the end of a pool with flowers as blue as the water, golden dewdrop wears its colors in an elaborate headdress atop a raucous plant. The branches take on a life of their own, arching here, drooping there, with hundreds of small leaves everywhere. Nickel-sized flowers form long clusters, called racemes, and open from the top down like water spilling from a pitcher. The flowers are essential to butterflies and other pollinating insects and attract hummingbirds in droves. Soon each flower becomes a gold berry, or drupe, transforming the flower racemes into chains of golden beads. Both flowers and drupes are breathtaking and long-lasting on the plants. Native to the West Indies and perhaps the Florida Keys, this plant has become established in Texas but is not considered invasive.

Plant Family
Verbenaceae

Other Common Name
None

Bloom Period and Seasonal Color
Blue and white flowers sporadically year-round in tropics, summer and fall elsewhere; gold berries

Mature Height × Spread
Up to 20 ft. × 6 ft.; averages 6 ft. × 3 ft. outside the tropics

When, Where, and How to Plant
Golden dewdrop is not invasive, but it will spread to drape its branches over the space allowed to it. Choose a site in sun or mostly sun with a reliable water source nearby. Limited exposure to salt spray will not damage it and it can tolerate regular breezes. Prepare a soil for garden beds or containers that has good drainage and is rich, organic, and fertile. Amend with organic matter such as ground bark and compost to achieve these goals. Space plants 3 feet apart in garden beds or plant one in a standard 12-inch clay pot. Grow it next to a warm wall and espalier the vigorous branches. Mulch beds and pots with 1 to 2 inches of organic material such as ground bark.

Growing Tips
Water and fertilize regularly and let the soil dry out only slightly between irrigations. Wilted plants may stunt and fail to bloom. Maintain organic mulch around the plants and work it into the soil as it rots naturally. Use a general-purpose garden fertilizer as often as directed from the spring through fall. Thin regularly to keep the plants in shape. Plan to root cuttings in the summer to grow indoors over the winter as young plants for the next year's garden.

Care and Propagation
Watch for slugs and snails on young plants. Propagate by tip cuttings taken from semi-hardwood and seeds.

Companion Planting and Design
Let it complement vines such as mandevillas and moonflowers. Grow golden dewdrop for contrasting flower and leaf shapes with yellow cestrums, Chinese red hat, and yellow bells.

Try These
Cultivars offer interesting features; for example, 'Alba' has white flowers while 'Variegata' has multicolored leaves.

Key Lime
Citrus aurantifolia

When, Where, and How to Plant

Where winters are frost-free, choose a site outdoors in full sun for key lime. Elsewhere, grow it indoors in a warm, sunny room when temperatures are below 40 degrees F and move it outdoors to a sunny deck whenever possible. Prepare a fertile, well-drained soil for key limes by adding sand and ground bark to potting mixes and garden soils. These plants need water regularly but cannot stand flooding or saturated soils. Space trees 2 to 3 feet apart in beds or provide a container 14 to 16 inches wide and deep. Take care to replant at the same level the tree was growing originally and soak deeply with water after planting. Mulch plantings in beds or pots with 1 inch of organic material such as ground bark.

Growing Tips

Water all citrus deeply and often. Let the soil dry out, but only slightly, between irrigations. Yellow leaves can be a symptom of water, fertilizer, or insect issues, or can indicate a need for lime in the soil. Fertilize four times annually with an all-purpose formula or a specialty fertilizer made for citrus. As the mulch decomposes, work it into the soil and replace it. Shrubby by nature, key limes can be trained into a small tree with selective pruning.

Care and Propagation

Whiteflies and mealybugs are common pests; be ready for them. Propagate by air layering, seeds, and rooting tip cuttings.

Companion Planting and Design

Choose companions that have strong contrast in leaf shapes and flower colors. Good choices are silver vase bromeliads, allamanda, and yellow bells. Add spineless yuccas and frangipani for textural variety.

Try These

Persian lime, or Tahitian lime, tastes sweeter and is slightly more cold hardy than key lime.

Among the most tender of the citrus plants, key lime had to be delicious and easy to grow or no one outside the tropics would know or appreciate its charms. It is both, and the world noticed. By the time the Spaniards brought key limes to Haiti in the early sixteenth century, this fruit was already an international traveler. Native to Southeast Asia and India, where it is still popular, key lime joined caravans to North Africa and the Near East with Arab traders, then on to Europe with the Crusaders. It is prized for its pale yellow, ripe fruit with juicy green pulp that lends itself to many culinary uses. The flowers are lightly fragrant and beautiful, five long white petals hidden inside an alluring purple bud. They shine like boutonnieres on the lapel of a bushy plant clad in a full-length overcoat of perky evergreen leaves.

Plant Family
Rutaceae

Other Common Name
Pie lime

Bloom Period and Seasonal Color
White flowers in the winter and spring

Mature Height × Spread
Up to 16 ft. × 4 ft.; averages 6 ft. × 2 ft.

Lady Palm

Rhapis excelsa

Lady palm conveys a properly elegant image for this dynamic tropical tree, but its other common name, bamboo palm, causes confusion. Another of many plants called bamboo palm, Chamaedorea seifrizii, truly looks like its namesake plant and deserves the moniker. Lady palm is a multistemmed tree with an angular, jaunty attitude. Smaller in stature than many palms, the Taiwanese native is the quintessential indoor palm. The lady has gusto, exploding in circular, shiny green leaves that look like hands with fingers spread wide in surprise. The brown or gray stems are stiffly erect, mature at about 1 inch in diameter, and spread by adding stems to the perimeter of the clump. The leaf whorls appear all along the stems in a staggered arrangement. Soon a thick stand can sport a dense mass of leaves, a bright green orb in the shade garden. Thinned to only a few stems, lady palm becomes a graceful, geometric wonder.

Plant Family
Arecaceae

Other Common Name
Bamboo palm

Bloom Period and Seasonal Color
Evergreen leaves

Mature Height × Spread
Up to 10 ft. × 10 ft.

When, Where, and How to Plant
Lady palm delights as a permanent garden plant anywhere temperatures do not dip below about 25 degrees F. for more than a few hours. It happily grows year-round as a container plant in any bright space, out of direct sun, but benefits from time outdoors in warm weather where it is not hardy. Prepare a soil that will be rich in organic matter, yet well drained. Amend native soils and potting mixes with organic matters such as compost and ground bark to meet those needs. Use one-third as much organic matter as other planting material. Space lady palms 5 feet apart in rows or provide a container large enough to accommodate the palm with plans to move it up to larger pots as necessary. Keep palms mulched with organic material.

Growing Tips
Lady palms are easygoing but cannot tolerate drought. Water and fertilize young plants very regularly. Soak the roots, and then allow the soil to dry out only slightly before watering again. Use a soluble fertilizer mixed in water monthly or apply a granular palm formula three times each year. Work rotting mulch into the soil in beds or pots and replace it promptly.

Care and Propagation
Few pests ever bother lady palms. Propagate by rooting suckers (new shoots from the roots).

Companion Planting and Design
Surround lady palms with Persian shield, rattlesnake plants, and impatiens in shades of pink and purple. Go for a large group with nun's orchids, dumb canes, philodendrons, and Malaysian orchids.

Try These
Wildly variegated forms of lady palm are popular in Japan and for gardens designed in the Japanese style.

Miniature Date Palm
Phoenix roebelenii

When, Where, and How to Plant
Miniature date palm grows best in bright sun, but it is well suited to dappled sun and even partial shade. This quality enables a potted palm to spend sunny summers by the pool and move into less light indoors without stress. Well-drained soil is essential, especially for container palms. Amend existing garden soils and potting mixes with ground bark and compost to ensure these needs are met. Space palms at least 4 feet apart in garden beds. A small plant in a huge pot will grow more roots than topside growth. Plant a miniature date palm in a container that accommodates its current size, with plans to move it up annually to a larger container. Water new plantings well and use a root stimulator fertilizer or compost tea if desired. Mulch lightly, if at all, with 1 inch of organic material.

Growing Tips
Miniature date palms grow slowly but benefit from a granular, complete, all-purpose fertilizer worked into the soil around the base of the plants four times each year. Repot annually as the palm is growing, or when needed to refresh soil of a mature planting. Root prune to repot in same container and maintain the palm's size. Water regularly with a slow, soaking drip.

Care and Propagation
There are few pests, but watch out for this palm's sharp spines. Propagate it by seed. See "Propagation" (page 36) for details.

Companion Planting and Design
Grow miniature date palms amid hibiscus and coral plants for accents, with gazanias and Cape daisies as groundcovers to complete the scene.

Try These
Edible date (*P. dactylifera*) is a close relative of miniature date palm and looks like a much larger version of it.

Miniature date palm is small compared to most palms, but its stately shape, dark good looks, and fine-textured canopy give it huge appeal. It is a single upright trunked tree by nature, dark gray-brown and covered with the remains of early leaves, only their fat bases left to mark where they grew. A single leaf can be 4 feet long and almost as wide at its maximum, full of lacy leaflets. They arise in a tight cluster and, taken together, a bunch of them is as frilly as any hat in the Easter parade. Male and female flowers are borne on separate trees, and the girls take the prize for showy flowers. They spill out in long chains, creamy orange beads that soon become black dates that shine like a jet necklace. Native to tropical forests of Southeast Asia, this easy-to-grow miniature date palm brings high style and strong impact to gardens.

Plant Family
Arecaceae

Other Common Name
Pygmy date palm

Bloom Period and Seasonal Color
Glossy medium green leaves

Mature Height × Spread
Up to 10 ft. × 6 ft.

Money Tree
Pachira aquatica

Native to Mexico and Central America, money tree is a magnificent wetland tree that, given enough water, will grow in uplands or containers as a small tree or large shrub. The base of the gray-brown trunks spreads out with clubby but strangely graceful, gnarled toes. The leaves grow to be large and glossy, reminiscent of schefflera, and are eaten as a vegetable when they are young and small. Nested inside the canopy, the blossoms' white petals fall back like booster rockets to unleash a fluffy bristle of white stamens tipped in fire engine red. They scream out to be pollinated and soon the reward appears as a prized edible. The pods are cocoa brown footballs that crack open in segments as if perforated to reveal amber nuts said to taste like peanuts. The nuts are eaten raw, fried, or roasted and then ground into flour for baking bread.

Plant Family
Malvaceae

Other Common Name
Guiana chestnut, wild cocoa

Bloom Period and Seasonal Color
Purple and red or yellow and pink flowers in the summer

Mature Height × Spread
Up to 40 ft. × 20 ft.; averages 6 ft. × 2 ft.

When, Where, and How to Plant
Plant it or place pots of money tree in sheltered locations, away from winds that can dry its leaves excessively. In pots or garden beds, these plants thrive in fertile soil made by amending existing soils and potting mixes with organic matters. Indoors, a sunny window can keep the plants growing, while a brightly lit room will keep them alive. When purchasing new specimens of this tree, select larger, older plants that can bloom sooner than young seedlings. Space the plants 3 feet apart in beds, or grow one in a standard 10-inch container. Keep new plantings watered well until they are established and have begun to grow. Mulch with 2 inches of organic material such as ground bark.

Growing Tips
Maintain a regular schedule of water and fertilizer for money trees, especially those being grown in pots. Do not allow the soil to dry out completely between irrigations. A few dropped leaves are expected, but heavy leaf drop usually indicates that conditions are too dry to sustain the plant. Keep plantings mulched to a depth of 2 inches in beds, 1 inch in pots; as organic mulch rots naturally, work it into the soil for additional nutrition.

Care and Propagation
Money tree has few pests to worry about. Propagate them by seed and cuttings.

Companion Planting and Design
Grow money trees with compatible plants that display diverse textures such as Brazilian red hot, papayas, lobster claw, and fragrant heliotrope. In dry environments, group a plant with other plants to raise the humidity around all of them.

Try These
Many money tree offerings are grown with braided trunks that have delightfully rough textures.

Navel Orange
Citrus sinensis

When, Where, and How to Plant
In frost-free areas, choose a site outdoors in full sun for orange trees. Elsewhere, grow this fruiting tree indoors in a warm, sunny room in the winter, letting it spend its summers outdoors on a sunny deck. Provide a fertile, well-drained soil; if necessary, add sand and ground bark to potting mixes and garden soils. Citrus trees need water regularly but they do not tolerate saturated soils. Space trees 2 to 3 feet apart in beds or provide a container 14 to 16 inches wide and deep. Note the place on the trunk where the desirable orange was grafted to its hardy rootstock (the graft union) and take special care to plant the tree at the same level it was growing originally. Water well and blanket plantings in beds or pots with 1 inch of organic material such as ground bark.

Growing Tips
Manage water wisely to grow the tastiest citrus. Water deeply, and then let the soil dry out, but only slightly, before irrigating again. Fertilize four times annually in early winter, late spring, summer, and fall, to correspond with a tree's blooming and fruiting cycles. Use an all-purpose formula or a specialty fertilizer made for citrus. As the organic mulch rots, work it into the soil around the plant and replace it.

Care and Propagation
Whiteflies and mealybugs are common pests, so keep your eyes open for them. Propagate by air layering, followed by grafting.

Companion Planting and Design
Choose companions with similar care needs and color variety such as Chinese violet, red justicia, and golden dewdrop. Grow your own edibles grove with key limes, plus curry and cinnamon trees.

Try These
'Hamlin' is a smaller tree, slightly hardier than the navel orange.

When people savor a taste of the navel orange, they become its ambassadors and will go to great lengths to share it with the world. Its nativity remains a mystery; perhaps it hails from China or India or someplace else, but its good flavor has taken navels far from home. Oranges found their way to the Mediterranean about 1500 and soon lent their name to the glass conservatories built to accommodate them, called orangeries. From there, oranges took the rest of the world by storm and today it is the most commonly grown fruit tree worldwide. An orange tree has one trunk and a dense, rounded crown of dark green leaves. Round fruits decorate the tree, glowing orbs full of vitamin C and potassium as nutritious as they are delicious. Though not the easiest tropical plant to grow—they are damaged by temperatures below 30 degrees F—navel oranges deserve all the attention they demand.

Plant Family
Rutaceae

Other Common Name
Orange tree

Bloom Period and Seasonal Color
White flowers in the winter and spring

Mature Height × Spread
Up to 20 ft. × 6 ft.; maintain in containers at 8 ft. × 4 ft.

Night-blooming Jasmine
Cestrum nocturnum

Night-blooming jasmine could be called the "romance plant," for all the courting couples it has sparked. One whiff of its distinctive, sweet yet musky fragrance and every garden becomes a seductive tropical paradise. Tall, erect stems are lined with shiny green leaves, long narrow ovals with distinctly marked midribs. One-inch-long white tubular flowers burst in clusters all over the plant and flare open at dusk. Their aroma penetrates the night air, guiding pollinator moths through the darkness to the nectar prize. Native to the West Indies, night-blooming jasmine thrives in containers, propagates readily, and may become a reliable garden perennial in the subtropics. Stroll the streets of the French Quarter in New Orleans on a late summer night for a total immersion experience in the dark charms of night-blooming jasmine. The flowers glow, the fragrance enchants, and everyone falls in love . . . with this plant.

Plant Family
Solanaceae

Other Common Name
Night-blooming cestrum

Bloom Period and Seasonal Color
Fragrant white flowers appear in the summer and fall

Mature Height × Spread
Up to 15 ft. × 5 ft.; averages 4 ft. × 3 ft. in beds and containers

When, Where, and How to Plant
Grow in places that are warm, sunny, and have access to water nearby. These plants cannot tolerate wet feet for more than a few hours but still need regular applications of water to thrive. Night-blooming jasmines can grow in dappled sun and still bloom, but dry soil stops their growth and flowering. Provide soil for beds and pots that is well drained, organic, and fertile. Amend as needed with organic matter such as compost and ground bark. Space 2 feet apart in beds or plant one in a container 14 to 16 inches wide and deep. Maintain a 1-inch layer of mulch around the plants. Lift plants grown in garden beds, prune them back by one third, and pot them up to spend the winter in an indoor garden.

Growing Tips
A simple regimen of water and fertilizer regularly applied enables night-blooming jasmines to grow large and bloom for months. Drip irrigation or soaker hoses will deliver more water into the root zone of this leafy plant than overhead sprinklers alone. During very dry times, water both ways to ensure good hydration and keep the leaves clean. Use a granular or soluble all-purpose fertilizer as directed until flowering begins, and then use it at half strength for the rest of the season outdoors.

Care and Propagation
Identify any "pests" before spraying! Watch for whiteflies, but also tiny white moths that do no damage. Propagate by cuttings in late summer.

Companion Planting and Design
Plant this cestrum with its cousin, yellow cestrum, for flowers day and night. Fill the front of a bed or a pot's edge with cat's whiskers and coleus for color and contrast.

Try These
Dwarf shrub jasmine (*Cestrum* sp.), has thicker, pastel purple tubular flowers.

Papaya
Carica papaya

When, Where, and How to Plant

Except in the wet tropics, pick a sunny site outdoors in warm weather and a bright space indoors to grow papayas. Choose Hawaiian papayas for gardens outside the tropics, and acquire a tree that produces male and female flowers, if possible, for better chances at producing fruit. Provide fertile, organic, well-drained soil amended with compost, ground bark, and composted manure to ensure its fertility and drainage. Whether in a garden bed or a container, the soil must be able hold its moisture between irrigation cycles without drying out or flooding. Grow one papaya tree in a container that is 12 inches across and deep. A papaya plant growing indoors reaches 2 feet tall in about one year. It can fruit the next year if it is transplanted at that height into the spring garden.

Growing Tips

Use drip irrigation or bubbler heads if needed to provide very regular water to papayas. To ensure a rapid growth rate, consider a timer on that water supply to deliver it regularly. Keep plantings mulched to a depth of 2 inches in beds and pots. Fertilize with an all-purpose granular formula four times each year and as organic mulch rots naturally, work it into the soil and replace it.

Care and Propagation

Few pests bother it in the home garden unless papaya fruit flies beset the plants. Propagate by seeds.

Companion Planting and Design

Let papaya's upright height set the stage among umbrella plants, star jasmines, and lobster claw. Or plant a pair of pots to bring rich greens to the design.

Try These

'Tainung' grows in containers and produces clusters of 3-pound fruits.

Not many plants have been grown in so many places for so long that no one can be sure of their origins, but papaya has become ubiquitous wherever the climate is sunny and hot with plenty of rainfall. Their stately shape, ornately cut leaves, and delectable, tropical taste are prized worldwide. The leaves grow as long as an arm and almost as wide in stunning green shades. Nowadays, they are divided into two main groups: the huge Mexican papaya trees with fruit to match and the smaller, more familiar Hawaiian papayas. They are a smaller version of the tree and bear the fruit most often seen in markets outside the tropics. Palmlike trunks are crowned with massive numbers of finely fingered, large leaves that shade the yellow-white flowers. Papaya flowers can be female, male, or both. The fruits are fleshy, sweetly aromatic, and spoon-tender when ripe. Together, they offer the epitome of tropical allure.

Plant Family
Caricaceae

Other Common Name
Tree melon

Bloom Period and Seasonal Color
Spring brings small whitish yellow flowers

Mature Height × Spread
Up to 20 ft. × 10 ft. for Mexican types; 8 ft. × 4 ft. for Hawaiian types

Sago Palm
Cycas revoluta

Long-lived and slow-growing, sago palms can tower as telephone pole-sized trunks with gigantic whirligig fronds held aloft. Younger, smaller, chunkier sago palms squat like toadstools in their containers and garden beds. At any size these round hunks of Nature are impressive with their stiff, arching, very uniform fronds in a perfect circle around the trunk. Each frond, the cycad equivalent of a leaf, begins tightly clinched and avocado green. Soon it unfurls with dark green leaflets in precise rows along each side, as symmetrical as the teeth on a comb. Sago palm does not flower and is either male or female. Older male plants develop prominent scaly cones, while females produce center clusters of modified leaves to hold golden spores and seeds. Native in a range from tropical Asia to Australia, sago palms are a staple of warm-weather landscapes worldwide and are a reliable houseplant companion elsewhere.

Plant Family
Cycadaceae

Other Common Name
Cycad palm

Bloom Period and Seasonal Color
Dark green fronds are evergreen in the tropics and subtropics

Mature Height × Spread
Up to 15 ft. × 5 ft. where hardy; up to 5 ft. × 5 ft. in pots

When, Where, and How to Plant
Select sites in sun, dappled sun, or part shade with a reliable water source nearby. These are not drought tolerant plants until they are well established and will stop growing or turn yellow in dry soils. Prepare a soil for beds and containers that is well drained, organic, and fertile. Use compost and other organic matters along with sand to amend soils and improve their structure. Space sago palms about 3 feet apart in rows or huge planters and plant small ones in pots 12 inches deep and wide. Plant at the same level it was growing originally and then soak deeply. Mulch with 2 inches of organic matter such as ground bark in beds but limit mulch in pots to 1 inch or less.

Growing Tips
Sago palms are easy to grow with a simple routine of regular water and occasional fertilizer. They cannot stand in water for long but need plenty of it regularly to produce and sustain thick trunks and heavy leaves. Fertilize sago palms with a soluble or granular all-purpose formula in the spring and summer. Maintain 2 inches of organic mulch such as ground bark around the base of a plant and work it into the soil as it decomposes.

Care and Propagation
Cycad scale is a big problem in Florida, but elsewhere few pests are problems. Propagate by removing suckers or the buds that form along the trunk.

Companion Planting and Design
Contrast a sago palm's fronds with bolder shapes of candlestick tree, night-blooming jasmines, and silver vase bromeliads in sunny sites. Give it green friends when you are growing it in part shade with selloum and caladiums.

Try These
In tall office atriums, queen sago (*C. circinalis*) delivers a huge tropical statement.

Spineless Yucca
Yucca elephantipes

When, Where, and How to Plant
Prepare a site for a spineless yucca that is sunny or mostly so where the soil is well drained and fertile. To improve both qualities in garden soil or potting mixes, amend them with organic matters such as compost or ground bark. Unlike most yuccas, spineless yucca is not drought tolerant and benefits from high overhead shade at midday in hot climates. These conditions produce fast growth, but the plants can also adapt to seaside conditions. Plant young yuccas close together to start a clump or space them 2 feet apart as rows. Provide a container no smaller than 12 inches deep and wide for one specimen. Maintain 1 inch of organic mulch around all plantings. Start yucca offsets in sandy, very well-drained potting mix to encourage rapid rooting.

Growing Tips
Spineless yuccas respond best to a growing regimen that allows the soil to barely dry out between watering. Use a granular, all-purpose garden fertilizer four times annually or use its soluble equivalent monthly. Work mulch into the soil as it rots for added nutrition and organic matter content. Repot container grown yuccas annually in pots that are deeper than they are tall to accommodate the plant's root system. Leave the suckers to encourage a clump or remove them for a singular, treelike effect.

Care and Propagation
Few pests bother spineless yuccas. Propagate by rooting suckers to gain more plants. See "Propagation" (page 36) for details.

Companion Planting and Design
Group spineless yucca's bold leaves with its equals—edible pineapples, devil's backbone, and miniature date palms. Grow some as a canopy above gazanias and gerber and Cape daisies.

Try These
Adam's needle, also called needle palm, is a hardy, trunkless yucca with blue-green spiked leaves.

Spineless yucca has a noble form, a grayish or light brown trunk that erupts with sword-shaped bright green leaves. They glow like a lighthouse beacon, refusing to be ignored, with dangerous-looking, sharply pointed tips. Their fierce façade is just that, however, an illusion undone by simply daring to touch the strangely supple, barbless leaves. The plants are upright, clumping, thick-trunked trees in their native environs in Central America yet they adapt readily, producing top growth in proportion to the root space available. Spineless yuccas are an excellent air cleaner and will not wilt unless extreme drought conditions prevail. It is a staple indoor plant that grows in single window offices and hotel atriums, and they tolerate such environments with characteristic aplomb. Spineless yucca is a long-lived celebration of essential tropical spirit—bold form, bright colors, and stalwart resilience in the face of changing conditions.

Plant Family
Asparagaceae

Other Common Name
Giant yucca

Bloom Period and Seasonal Color
Evergreen leaves; spikes of white, bell-shaped flowers in summer.

Mature Height × Spread
15 ft. to 30 ft. × 10 ft.; averages 6 ft. × 2 ft.

Weeping Fig
Ficus benjamina

In the tropics, critters of all sizes and shape have four specific needs: food, water, nests, and safe resting places. Some plants fill every need while others specialize in one aspect of habitat and do it well. Such a plant is weeping fig, evergreen and luscious, full of narrow crevices and oddly curved leaf tips that are ideal camouflage for the tiniest pollinators. The tree is native to tropical Asia and indigenous to a huge swath from India to Australia. It usually has one trunk and soon branches into an open, or banyan, habit. Each branch ends in slender twigs that droop slightly, holding the leaves in a delicate balance that lets the leaves hang, or weep, distinctly. Mature trees are (somewhat) drought tolerant when established in shady sites. Grayish white bark and glossy dark green leaves with lighter undersides mark weeping fig. It sets an uptown tone wherever it is grown.

Plant Family
Moraceae

Other Common Name
Laurel fig, Cuban laurel

Bloom Period and Seasonal Color
Shiny green leaves year-round

Mature Height × Spread
Up to 60 ft. × 20 ft. in the tropics; up to 15 ft. × 3 ft. in containers

When, Where, and How to Plant
Weeping figs are a staple landscape plant in the tropics wherever there is bright light without direct sun and soil conditions are moderate. It grows large in its native climates, with root systems that can lift a sidewalk. Elsewhere, find or create a similar site for this long-lived plant and prepare a soil that is organic, fertile, and well drained. Amend garden soils and potting mixes as needed to achieve these conditions. Weeping fig plants are available in variously sized pots and will continue to grow until their roots become crowded. Mulch weeping figs very lightly with organic material, no more than 1 inch in garden beds or pots. To ease the transition to wintering indoors, bring weeping figs into the house in early fall. Give the plants a spot away from doorways and heater vents (to avoid drafts).

Growing Tips
It is essential that weeping fig's soil be allowed to dry out slightly between waterings because saturated roots will rot. The plants are not truly drought tolerant, and will drop leaves if stressed in either direction. Fertilize monthly with a soluble general-purpose formula while weeping figs are young. Mature trees benefit from slow-release formulas. Container growers should repot annually, moving up in container size, until the tree reaches desired height, then root prune and refresh soil in same pot each year.

Care and Propagation
Watch for aphids on new growth and scale insects on old wood. Propagate by cuttings.

Companion Planting and Design
Differing plant heights can create a stairstep design, from weeping figs down to scheffleras and then to collections of peperomias. Contrast a weeping fig's glossy green with the jade plant's more yellow-green tones and snake plant's patterned leaves.

Try These
Look for variegated laurel figs and rubber trees (F. elastica) and lyre leaf fig (F. lyrata).

Yellow Bells

Tecoma stans

When, Where, and How to Plant

Choose a sunny site for yellow bells that has a reliable water source nearby. The plants are not drought tolerant until they are mature and well established; neither can they stand a saturated soil. They can happily spend the spring, summer, and fall outdoors in temperate zones and transition indoors for the winter as saved seeds and rooted cuttings. Prepare a soil for beds or containers that is well drained as well as fertile and organic. Amend as necessary with organic matters such as compost to create the moderate conditions these plants prefer. Plant outdoors anytime of year in the tropics, in early spring elsewhere. Space young plants 18 to 24 inches apart or provide a standard 10-inch pot for each plant. Mulch with 2 inches of organic material such as ground bark.

Growing Tips

Yellow bells are best treated as creatures of habit, watered very regularly so the soil scarcely dries out between irrigations. Fertilize regularly by working a slow-release formula into the soil and supplementing that with a soluble mixed into the water every other month. Let seedpods dry on the plant. Start seeds in late winter or carry rooted cuttings over the winter for your next crop of yellow bells.

Care and Propagation

Few pests ever bother yellow bells. Propagate by seeds and cuttings taken from semi-hardwood in the summer.

Companion Planting and Design

Let yellow bells be the star surrounded by Persian shield, bat face cupheas, scarlet sages, and cat's whiskers. Grow a plant with angel trumpets and horn of plenty to continue the trumpet-flower theme.

Try These

'Gold Star Esperanza' blooms early and often, while 'Orange Jubilee' and 'Burnt Out' have orange flowers.

Indigenous to the American tropics from Argentina and Brazil to south Florida, this woody shrub or small tree occupies an important, though less well-known, branch of its family tree. Related to catalpa and trumpet vine, yellow bells' gracious flowers do not so much resemble their relatives as mimic them. All share a lower petal that is a bit longer than the rest and thus looks pouty. But the cousins have flower clusters that scream out, brassy jazz bands to the carillon sounds orchestrated by yellow bells. These flowers form a whirligig of cheerful, friendly bunches set in bright green, toothy leaflets. The plants offer strong rounded form and a long bloom season that makes them a welcome tropical plant for every garden during the summer. Yellow bells is a stalwart plant outdoors and remains friendly all year as a welcome addition to an indoor garden.

Plant Family
Bignoniaceae

Other Common Name
Yellow elder

Bloom Period and Seasonal Color
Gold trumpets year-round in the tropics, spring and fall elsewhere

Mature Height × Spread
Up to 20 ft. × 5 ft.; averages 5 ft. × 3 ft. outside the tropics

Yellow Gardenia
Gardenia carinata

Yellow gardenias can seem a bit snooty, always looking so perfect even when they are not in bloom, always insisting that evenings be cooler than days. If perfect patent leather leaves and seductive aroma qualify as elite, then yellow gardenia makes the A list. The leaves are a class act: stalwart, evergreen, deeply veined, and lighter on their undersides. In tropical terms, yellow gardenia is debutante coy, not covered with flowers like most of its mates. Instead the blooms dot the plant, all the more precious for their scarcity. Proper pinwheels as big as half dollars open in creamy tones that soon change to warm, deep yellows with orange overtones. The flowers are spicy with fragrance and open a few at a time so their colors are displayed for weeks. Sweet enough to take home to Mama, yellow gardenias can be a long-lived container plant outside the tropics.

Plant Family
Rubiaceae

Other Common Name
Golden gardenia

Bloom Period and Seasonal Color
Yellow and yellow-red primarily in the spring and summer

Mature Height × Spread
Up to 30 ft. × 15 ft.; up to 8 ft. × 3 ft. in containers

When, Where, and How to Plant
Yellow gardenia is a prized container plant for late winter flowers indoors everywhere that thrives when evening temperatures are 10 to 15 degrees cooler than days. Choose a sunny site for yellow gardenias, which can grow outdoors year-round in the warmest subtropics as well as in its native tropics. This plant grows best in soil that is organic, fertile, and well drained. It is not drought tolerant nor is it able to withstand flooding. Prepare a soil in beds or containers that combines garden soil or potting mix with a variety of organic matters to achieve these conditions. Select a container that is 3 to 4 inches wider all around than the base of the plant, preferably a clay pot for superior drainage.

Growing Tips
Give yellow gardenias a regular routine that allows the plants to dry out slightly between waterings. Use a granular, complete, all-purpose fertilizer in the soil around the plants four times annually or use an equivalent soluble formula six times a year. Keep mulch to a minimum in beds, no more than 1 inch of organic material. Prune annually after flowering as much as needed to control size and shape. Repot container plants at the same time if more space is needed for that year's growth.

Care and Propagation
Watch out for mealybugs and whiteflies. If you like, yellow gardenias can be propagated by cuttings.

Companion Planting and Design
Grow several flowering shrubs with yellow gardenias, including hibiscus and yesterday, today, and tomorrow plants. Combine them with kanga paw, Cape daisies, and gazanias in dry beds. Surround yellow gardenias with gerber daisies.

Try These
G. radicans, dwarf gardenia, is almost a groundcover with evergreen leaves and fragrant white flowers.

Yesterday, Today, and Tomorrow

Brunfelsia pauciflora

When, Where, and How to Plant

Yesterday, today, and tomorrow can spend the warm months outdoors and readily makes a smooth transition to a bright room indoors for the winter. Outdoors, in beds or pots, a site in bright shade, morning sun, or dappled sun will bring on sturdy growth and plentiful flowers. Provide a soil that has excellent water-holding capacity, yet still drains well. Garden soils and potting mixes will need amendments of compost and composted manure or similar materials. Plant one in a 5-gallon container or space them 2 feet apart in beds and mulch deeply. Pot up garden grown plants in the fall and grow indoors over the winter. This plant can also be overwintered in a shed or anywhere its roots will not freeze, but flowers may bloom late the following year.

Growing Tips

Maintain a consistently damp soil with regular additions of compost. Fertilize the plant with a complete all-purpose formula in the spring and summer. Work in organic mulch as the organic matter decomposes and topdress a plant's base with composted manure in late spring. Control the plant's ultimate height by top pruning after it flowers and by root pruning when repotting. Left unpruned, plants will be more upright and produce many seedpods.

Care and Propagation

There are no pests of note. Propagate by tip cuttings in late spring or summer from semi-hardwood.

Companion Planting and Design

Grow pots of yesterday, today, and tomorrow with colorful skirts of impatiens, begonias, or Persian shield. Plant a mixed summer hedge with canna lilies and chocolate plants.

Try These

B. australis has somewhat greater cold tolerance than *B. pauciflora*.

Surely there is romance in a plant whose name sends a message of loyal and enduring love. Those who grow it certainly develop great affection for this quiet plant's reliable presence in the garden. Simple green leaves cover a shrubby plant to make a perfect backdrop for the endearing, unusual flowers. Late spring brings on scores of fluffy purple scalloped blossoms with a charming habit that is rare among flowers. They change color, not over weeks as the flowers mature, but in the course of three nights, first to lavender and then to white! Soon the plant looks as if a polka dot tablecloth was dropped over it, ready for a picnic basket and wineglasses. Native to Brazil, this plant likely evolved its color-changing habit as most plants do: to attract certain pollinators or deter predators. Its natural defenses are a gardener's valentine.

Plant Family
Solanaceae

Other Common Name
Morning, noon, and night

Bloom Period and Seasonal Color
Spring brings purple, lavender and white flowers simultaneously

Mature Height × Spread
Up to 6 ft. × 4 ft.; averages 3 ft. × 3 ft. in pots everywhere and in subtropical beds

Vining Tropicals

Bougainvillea
Bougainvillea spp.

Bougainvillea brings gasps and broad smiles at first glance, whether in tropical environs or your local garden center. It is difficult to imagine so many flowers and so much color on one plant, but bougainvillea delivers the show for months. Its leaves are longer than wide, colored dark green or variegated with cream markings. Handsome branches arch out and down stiffly, all the better to display the profuse numbers of flowers. They are clusters of cupped bracts in red, pink, purple, orange, or yellow, embracing tiny white flowers. These bracts look fragile, but they are not. This spectacular plant was named to honor Admiral Louis-Antoine de Bougainville, commander of the ship that carried the first Europeans who ever saw its glory. The magnificent performer Carmen Miranda was known as "The Brazilian Bombshell," but to gardeners, bougainvillea plants in full bloom present stiff competition for that title.

Plant Family
Nyctaginaceae

Other Common Name
None

Bloom Period and Seasonal Color
Papery bracts red, pink, purple, orange, and yellow throughout the year (mostly summer)

Mature Height × Spread
Vine trails to 20 ft. in the tropics; averages 5 ft. to 10 ft. in containers (pruned)

When, Where, and How to Plant
Bougainvillea grows and flowers best in very well-drained, even sandy soil. Where this condition cannot be met, it is best grown in containers using a prepared potting mix. Products intended to maintain moisture in the mix may stay too wet unless you amend them with sand and ground bark. Plant three small bougainvilleas in a 10-inch basket or one in a mixed container with other plants best grown on the dry side. Provide full sun and elevated spaces for these stiff vines to trail without interruption. Space plants for excellent air circulation around each one. Choose hanging baskets, or place pots on columns or brick walls for most exposure to sun and heat in temperate climates. Move them indoors before temperatures drop below 50 degrees F.

Growing Tips
Let bougainvilleas dry out thoroughly between waterings when they are not in bloom. When flowers appear, water the plants regularly but not excessively. Lightly prune stem ends occasionally to encourage blooming. Delay repotting, as there will be many more flowers on rootbound plants. Fertilize bougainvillea regularly but sparingly. Use a soluble flower formula every other month. Move pots into a sunny, warm room for the winter. Cut back large plants to lessen leaf drop during this transition.

Care and Propagation
Watch for mealybugs and aphids on tender new growth. Propagate by cuttings if you want more plants. See "Propagation" (page 36) for details.

Companion Planting and Design
Bougainvilleas bring their own heat to the landscape and combine well in hot, dry gardens with desert rose, chameleon plants, spineless yuccas, and jade plants.

Try These
Try 'Raspberry Ice' for fuchsia flowers and variegated leaves and 'Scarlett O'Hara' for its red bracts.

Chinese Violet
Telosma cordata

When, Where, and How to Plant
Select a site in full or part sun away from winds and salt spray. The plants can grow outside in warm weather and readily adapt to indoor culture for winter outside the tropics. Prepare a soil that is organic, fertile, and well drained. Amend existing garden soils and potting mixes as needed with compost and other organic matters to achieve these conditions. Space plants 1 foot apart in beds and mixed planters, or provide a standard clay pot with a diameter of 8 to 10 inches. Provide modest support such as a 3-foot-tall wire or bamboo pole. Clay or slatted wood containers make it harder to overwater this plant. Water new plantings well but avoid saturating the soil. Mulch pots and beds with 1 inch of organic material such as ground bark.

Growing Tips
Water very regularly, and let the soil dry out only a bit between waterings in the growing season. Water even less in the winter, indoors or out, when the plants will go semi-dormant and roots can rot. Fertilize regularly during the growing season with a granular or soluble product. As mulch rots, work it into the soil and replace with fresh organic material. Prune Chinese violets in late winter as new growth is emerging.

Care and Propagation
Watch for spider mites in dry weather and remember that desirable caterpillars may strip the foliage; identify before you spray. Chinese violets can be propagated by cuttings and seeds.

Companion Planting and Design
Grow Chinese violets as a centerpiece with coleus and licorice plants for trilevel interest. Pair them with Chinese hat plants for astounding summer flowers, and let them shine in front of hyacinth beans or moonflowers.

Try These
You cannot go wrong by planting the species.

Chinese violet is a small but muscular vine related to the Asclepias *genus, known for their milky sap and irresistible nectar. Members of this family host butterflies and sustain diverse pollinators in their native environs. Chinese violet flowers look like their kin but are thicker, almost armored when compared to the temperate zone milkweeds. From these golden yellow starbursts comes an unmistakable, almost cloying sweet fragrance. The flowers seem to be made of thick paper and are borne in clusters on thick stems all along the stumpy vines. This is a workmanlike plant covered in fat, heart-shaped leaves with distinct veins that define their space like six-pack abs. It pumps out a succession of bloom clusters for months, fists full of cream-throated, golden fragrance machines. Chinese violet slows its growth in the fall and transitions readily to a cool, dry room indoors where necessary.*

Plant Family
Apocynaceae

Other Common Name
Tonkin creeper

Bloom Period and Seasonal Color
Fragrant yellow flowers from the spring through fall

Mature Height × Spread
Up to 3 ft. × 2 ft.

Corkscrew Flower
Vigna caracalla

At first glance, corkscrew flower vine looks like its common garden relatives, pole beans, with broad, dark green leaves that cover strong stems. But instead of producing haricot verts for the dinner table, these vines deliver flowers that can best be described as complex and exotic. If pole bean flowers are a song, corkscrew flowers are a symphony performed by a full-out, world-class orchestra. Extraordinarily fragrant, these flowers present first in a cluster of white buds that look like ivory snails crawling on the dark leaves. Moving from the bottom up on the cluster, each bud uncoils to reveal pinkish, purplish layers of baffles that protect the entrance with an enchanting maze of confusing geometry. The flowers twist and turn in on themselves in marvelous contortions. The purpose of these ingenious complications is to direct pollinators to enter a flower's chambers only on the left and thus prevent contamination.

Plant Family
Fabaceae/Leguminosae

Other Common Name
Corkscrew bean

Bloom Period and Seasonal Color
White, lavender, pink, and yellow flowers sporadically all year

Mature Height × Spread
Up to 20 ft. vine; averages 10 ft.

When, Where, and How to Plant
Choose a sunny site for this climbing legume and set up support for its vines such as a trellis, bean teepee, or dead tree. Provide a soil that is both rich in organic matter and well drained. Amend existing garden soils or potting mixes with ground bark or sand to achieve those conditions. Plant anytime in the tropics, or start seeds in peat cups or pellets indoors in late winter to move outside in warm weather. Space vines 6 to 8 inches apart along an 8-foot-tall trellis. Be sure none of the peat cup stands above the soil line at transplant. The peat pot will act like a wick and draw water away from the plants' roots. Provide cotton string for the vines to grab first if supports are not at ground level.

Growing Tips
Provide a schedule of regular watering with time in between to let the soil dry out somewhat before watering again. Spray water on vines occasionally in dry weather. Use a complete formula fertilizer made for garden plants occasionally from the spring through fall. Maintain 1 inch of organic mulch around a plant's base and work it into the soil as it rots.

Care and Propagation
Watch for spider mites in very dry weather. To propagate, root cuttings from late summer growth and/or save seeds to continue corkscrew flower from year to year outside the tropics.

Companion Planting and Design
Plant a friendly group in front of corkscrew flowers with musical note plants, devil's backbone, and ti plants. Edge a bed or planter with China asters or gerber daisies for early season color.

Try These
Grow corkscrew flowers with another tropical legume, hyacinth beans (*Dolichos purpureus*).

Flaming Glorybower
Clerodendrum splendens

When, Where, and How to Plant

Flaming glorybowers thrive in an organic, fertile, well-drained soil. In garden beds and containers, the soil must have a structure that enables it to be watered very regularly and remain moist between irrigations, yet still continue to be well drained. Amend garden soils and potting mixes with organic matters such as compost to achieve this condition. To allow room for suckers to thicken the stand, space plants 4 feet apart or grow in containers 12 to 14 inches wide and as deep. Flowers are more abundant in a sunny location, but the vine benefits from late afternoon shade in the summer. Provide shelter from strong winds and salt spray, and install a trellis when planting or repotting flaming glorybowers. Mulch plantings with 2 inches of organic material such as ground bark.

Growing Tips

Determine a reliable watering plan such as using soaker hoses to meet its need for very regular applications of water. Fertilize glorybowers three or four times annually using a complete all-purpose formula. Choose a product with some slow-release action in the summer. As mulch decomposes, work it into the soil to increase its organic matter content, and replace with new mulch. Glorybowers can be a bully. Prune vines to keep them tidy and contained on their trellises.

Care and Propagation

You will not have to worry about many pests, but watch out for opportunistic aphids. Propagate by rooting softwood cuttings or suckers. See "Propagation" (page 36) for details.

Companion Planting and Design

Keep the red-orange heat going all year in a design combining it with lobster claw and calico plants. Add fragrant heliotrope and star jasmines for summer contrasts and sweet aromas.

Try These

Other *Clerodendrum* plants worth growing include butterfly flower and harlequin glorybower.

An orchestra of cupped orange tubas sings in dense, wide clusters at the tips of this twining vine. Each flower is a stocky, short tube that pops open with the colorful petals and prominent stamens that curve upward. Gathered together in such huge bunches, flaming glorybower can, and does, stop traffic. Yet for a plant so huge visually, flaming glorybower is quite well behaved and gardener-friendly. The vine is small compared to many tropicals, somewhat woody, and easily grown in a container outside the tropics with a sturdy 6- to 8-foot-tall trellis. Vigorous, with oblong leaves as long as a hand, the vine grows up and spills out all at once, and it blooms equally well growing up or sprawling down. Flaming glorybower adjusts to indoor conditions well and blooms in a sunny room or in a greenhouse as it does in the wild—in late winter to herald the coming of spring.

Plant Family
Lamiaceae

Other Common Name
Scarlet glorybower

Bloom Period and Seasonal Color
Red to red-orange flowers in late winter or spring

Mature Height × Spread
Up to 12 ft. woody vine

Flowering Pandanus
Freycinetia cumingiana

Native to the Philippine Islands, flowering pandanus grows slowly to great heights yet only a few ever reach direct sun. This bushy, viney plant has slender trunks seldom more than 1 inch around and vigorous. They sprout long stems lined with leaves shaped like lilies that drape gracefully like wimpy palm fronds over whatever is near. They wander unless pruned, as their trunks send out aerial roots reaching to grab available supports. For such a nomad in nature, they readily take direction and never bully other plants. Flowers that simply shout "tropical!" appear at the ends of each stem beginning in late fall and continuing into the spring. Each is a creamy orb of orange bracts that opens gradually to reveal a lush, perfect cone of flowers at its heart. Flowering pandanus looks good enough to eat. The rumor is the flowers not only look rich enough to eat, they taste sweet too.

Plant Family
Pandanaceae

Other Common Name
Climbing screwpine

Bloom Period and Seasonal Color
Deep orange, red, and pink flowers from the fall through spring

Mature Height × Spread
Up to 20 ft. vine; averages 8 ft. in containers

When, Where, and How to Plant
These plants flourish in garden soil in warm weather and can be lifted to overwinter as potted plants in the temperate zone. However, outside the tropics, flowering pandanus is usually grown as a container plant because it blooms in late winter and early spring. This plant is not drought tolerant, nor can it stand flooding for more than a few hours. Choose a site that is mostly sunny with a reliable water source nearby and consider a timer for greater reliability in its use. Prepare a soil in garden beds or containers that is richly organic and fertile with very good drainage. Put one flowering pandanus in a standard 12-inch pot or space plants 12 inches apart in beds. Mulch beds and pots with 1 to 2 inches of organic material.

Growing Tips
Water flowering pandanus plants deeply on a regular schedule and let the soil dry out only slightly between irrigations. Maintain organic mulch around the plants and work it into the soil as it rots naturally. Mix a general-purpose garden fertilizer into the water monthly or use a granular, slow-release formula four times annually. If plants do not bloom after a year, change to a formula made for flowering plants. Prune in the spring to control growth, if necessary.

Care and Propagation
Few pests bother flowering pandanus. Propagate more by suckers, cuttings, and seed.

Companion Planting and Design
Provide support for a plant and surround its base for contrast with coleus and polka dot plants. Grow flowering pandanus up a tree and flank it with yellow bells and copperleafs for complementary colors.

Try These
F. arborea, as its species name suggests, grows in trees with crazy spirals of leaves and purple flowers.

Glory Lily
Gloriosa superba

When, Where, and How to Plant

Choose a site in sun or part sun that has a reliable water source nearby. Prepare the soil so that it is organic, fertile, and well drained. Plant glory lilies next to another plant (or trellis) that can support its thin vines without crowding, such as 2 feet apart under an evergreen hedge. One glory lily plant can soon cover a small teepee trellis in a large mixed container. These plants thrive in humid, warm weather, and can be grown through the winter as potted plants in the temperate zone. Keep beds and pots mulched with 1 to 2 inches of organic material such as ground bark. Indoors, provide bright light and humidity, as with ferns. Put plants on a tray of gravel and mist both the plants and tray frequently to increase humidity.

Growing Tips

Water and fertilize often enough so the plants dry out only slightly between irrigations. Use a general-purpose fertilizer, either a soluble or granular formula, regularly from the spring through fall. Dig up tubers after flowering and repot or replant them right away to promote future flowers. Glory lily may perennialize in the subtropics, where it will die back and return from its roots each spring. Mulch outdoor plants well in the fall but do not allow them to become waterlogged. Wear gloves when handling glory lily rhizomes and prune only when you want the plant to die back.

Care and Propagation

Glory lilies are not bothered by pests. Propagate by removing offsets from around the base of a plant and by dividing the tubers.

Companion Planting and Design

Plant with sturdier vines such as Chinese violets, moonflowers, or hyacinth beans that can support another vine without compromising their own. Flank a planting with a warm wall or other shelter for faster flowering.

Try These

'Nana' is a dwarf version of the gloriosa lily, while 'Lutea' has solid yellow flowers.

Gardeners work to keep vines from climbing over shrubs and into trees, but this plant calls for an exception to that rule. Native to Africa, glory lily is a wimpy vine compared to other tropical climbers, which allows it to coexist with other plants for support. It is a true lily, seen in its underground tuber and leaf shape. But these leaves form tendrils at the leaf tips that enable the plant to grab hold of whatever is nearby. Glory lily flowers are spectacular, animated and oddly graceful from start to finish. Pale green buds curve down in a ploy to go unnoticed until the crinkly petals can stretch out and arch backward into boisterous red and yellow flames. As if the petals need an easel, stamens splay out beneath them in an unusual display of support. Like the well-mannered vine it is, glory lily dies back neatly after it blooms.

Plant Family
Colchicaceae

Other Common Name
Tiger claw lily, gloriosa lily

Bloom Period and Seasonal Color
Yellow and red twining flowers

Mature Height × Spread
Up to 8 ft. × 3 ft.; averages 5 ft. × 1 ft.

Golden Trumpet
Allamanda cathartica

Few tropical vines are as spectacular in bloom as golden trumpet, and none is easier to maintain. Glossy medium green leaves are 3 to 4 inches long and climb high, their surfaces so slick they look polished. Fuzzy brown buds soon appear all along the vines and give the cultivar 'Hendersonii' its common name, brown bud allamanda. The warm cocoa buds open to reveal sunny yellow, fluted trumpet flowers as wide as their leaves are long, with throats almost as deep. The flowers are waxy, fragrant, and plentiful from the spring through fall. Grown as a vine or pruned into a bushy shrub, allamanda blooms on new growth. Although it is technically native to Brazil, its popularity extends worldwide in the tropics. Named for Dr. Frederick Allamanda, a nineteenth-century professor of natural history, the vine's name is also said to mean "heavenly chief." For such a romantic and optimistic plant, there might be no better moniker.

Plant Family
Apocynaceae

Other Common Name
Allamanda, brown bud allamanda

Bloom Period and Seasonal Color
Fragrant yellow tubular flowers appear spring through fall if grown as an annual

Mature Height × Spread
Vine up to 50 ft. × 3 ft.; averages to 10 ft.

When, Where, and How to Plant
Golden trumpets can be planted outdoors in warm garden soil or will happily reside in a 12- to 15-inch container. Provide a richly organic, fertile planting soil or container growing mix and a location in full or part sun. Regular watering is essential to growing this plant, but the need is less pressing with less sun. Allamanda appreciates heat from a warm wall to radiate onto its vines, but its roots benefit from a cooler environment. Keep the base mulched, even in pots, with 2 inches of organic matter and use plants alongside it to shade the root zone. Train unpruned vines to twine onto a trellis or around any skinny pole. *All parts of golden trumpets are poisonous if eaten; avoid planting where it could become a hazard.*

Growing Tips
Water and fertilize the vines regularly during the growing season. Keep organic mulch around the base of a plant and use a complete soluble formula fertilizer twice each month. This easy-to-grow vine twines readily but responds well to pruning. Remove seedpods to keep new growth and flowers coming. Golden trumpet drops its leaves when plants suffer water stress. Yellowing between leaf veins, or chlorosis, indicates a need for more organic matter in the soil.

Care and Propagation
There are few pest problems to bother this plant. Propagate from 4-inch tip cuttings in summer. To reduce time indoors over the winter, root hardwood cuttings in late fall.

Companion Planting and Design
Golden trumpet vines and mandevilla vines quickly climb opposite columns to sweetly frame a doorway. Combine in pots or beds with chenille plants, bird of paradise, or night-blooming jasmines.

Try These
'Williamsii' is a smaller plant overall with semi-double, delicately fragrant flowers.

Hyacinth Bean
Dolichos purpureus

When, Where, and How to Plant

In frost-free climates, hyacinth beans will be perennial or reseed itself each year. Elsewhere, start seeds indoors four weeks before your area's last frost is expected or purchase small plants in the spring to grow outdoors as an annual. Choose a site in sun with a reliable water source nearby. Prepare a soil in garden beds that is richly organic and fertile with very good drainage. Space plants 6 inches apart along a sturdy support. Water new plants well and begin to train vines to their trellis as soon as possible. To help them grab hold, tie cotton string to the bottom of the trellis in enough length to reach the plants. Coax the vine onto the string first. Mulch beds with 2 inches of organic material.

Growing Tips

Like its edible relatives, hyacinth beans need water very regularly. Consider soaker hoses for outdoor plantings and a timer on that water supply to deliver it regularly, particularly in the subtropics. Keep plantings mulched to a depth of 2 inches in garden beds. As organic mulch rots naturally, work it into the soil for to maintain adequate levels of nutrition. Vines that do not bloom need less nitrogen fertilizer or more sun.

Care and Propagation

It does not suffer from many pests. Propagate hyacinth beans by planting seed in January, or six weeks before last frost in your area. See "Propagation" (page 36) for details.

Companion Planting and Design

For fast shade and months of flowers, install a pergola or an array of trellises with hyacinth beans, moonflowers, Chinese violets, and Mexican flame vines.

Try These

Emerald creeper is another leguminous vine with impressive stature and pods.

Native to Vietnam, where it is called lablab, hyacinth bean is a fine example of tropical design qualities in a vigorous but well-behaved vine. It demands attention for its rugged good looks and fast growth rate, but it is easily trained onto a trellis or dead tree and never bullies other plants. Fat seeds sprout into thick ropes of stems covered with bulky, coarse-looking leaves like garden beans on steroids. Hyacinth bean stems and leaves grow rapidly, deep green with purplish overtones that shimmer as the sun moves over the vines. The blooms are shaped like traditional legume flowers, mostly in hot shades of purple with pink and even white flower parts on others. The effect is that of a floriferous chandelier that lights up the summer garden. Soon after it flowers, shiny purple seedpods form, which are full of seeds for the next crop of hyacinth beans.

Plant Family
Fabaceae/Leguminosae

Other Common Name
Lablab

Bloom Period and Seasonal Color
Green leaves with light purple flowers in the summer followed by dark purple bean pods

Mature Height × Spread
Up to 15 ft. vine

Madagascar Jasmine
Stephanotis floribunda

Madagascar jasmine flowers can solve one of life's biggest little problems: how to put a smile on a nervous bride. Just the sight of them in her wedding bouquet—not to mention their clean, calming aroma—does the trick, and it's no wonder. Smooth, deep green leaves cover coarse brown vines to create a stiff canopy that is perfect to show off the bloom clusters. The waxy white flowers are slick tubes that burst open into the surprisingly fragrant five-pointed stars. Madagascar jasmine plants grow slowly into small but stocky evergreen vines in the tropics, and need no more than a 4-foot-tall support to grow in containers elsewhere. Cultivated for the florist trade and in gardens, greenhouses, and sunrooms everywhere, Madagascar jasmine has come to be seen as a symbol of marital bliss said to last long after the Big Day.

Plant Family
Apocynaceae

Other Common Name
Bride's veil jasmine

Bloom Period and Seasonal Color
White flowers all year, primarily in the summer

Mature Height × Spread
Up to 30 ft. vine; averages 6 ft.

When, Where, and How to Plant
Choose a sunny location near a reliable source of water to meet the needs of this jasmine. The plants can grow well outdoors in a bed and then be potted up for the winter, or be grown year-round in pots. Prepare a soil for beds and pots that is richly organic, fertile, and well drained. Add organic matter such as compost to existing garden soils and potting mixes to ensure these qualities. Space plants 12 to 14 inches apart in subtropical zone garden beds and, several feet apart in tropical zone ones, or use a pot at least 10 inches deep and wide for one plant. Clay pots offer the advantage of extra drainage through their surfaces. Do not let new plants dry out before they become established, and water deeply to encourage deep rooting.

Growing Tips
Madagascar jasmines require plenty of water, but they do not tolerate a saturated root zone. Leaves will drop and flowering will stop in reaction to extremes of drought or flooding. Let the soil dry out, but only slightly, between waterings. Select a general-purpose garden fertilizer with a complete, balanced formula and use it four times each year. Keep a mulch blanket around the base of the plants and work it into the soil as it rots.

Care and Propagation
Be prepared to fight off mealybugs. They can be propagated by seeds and cuttings.

Companion Planting and Design
Let madagascar jasmines take center stage with a chorus of Persian shield and polka dot plants for a pink and white scene. Or train it on an obelisk surrounded by the deep greens of sago palms, African gardenias, and allamanda.

Try These
If this aroma does not please the bride, try another traditional bouquet flower of orange blossom.

Mandevilla Vine
Mandevilla spp.

When, Where, and How to Plant
Create a soil for garden beds or containers of mandevilla vines that is richly organic and fertile with very good drainage. Amend existing garden soils and potting mixes with ground bark and compost to meet this vine's needs. Select a sunny site near a reliable water source and install a 6- to 8-foot-tall trellis. Plant outdoors after all danger of frost has passed, or during the rainy season in frost-free zones. Space plants 8 inches apart or grow two in a standard 12-inch pot. Water new plantings well and mulch with 2 inches of organic material. Mandevilla vines thrive in garden soil year-round in frost-free zones and can be cut back, dug up, and potted to spend the winter as a container plant where necessary. It may be hardy to the roots in protected areas outside the tropics.

Growing Tips
The key to keeping mandevilla vines in bloom is maintaining a simple schedule of water and fertilizer. Irrigate plants regularly and let them dry out only slightly between irrigations. As the mulch blanket around the plants rots, work it in to the soil and replace it promptly with fresh organic material. Use a general-purpose garden fertilizer as often as directed from the spring through fall when grown as an annual, or year-round in the tropics or as a container plant.

Care and Propagation
Watch for aphids on new growth and flower buds. Propagate by cuttings taken in the summer.

Companion Planting and Design
Let mandevilla vines grow up lampposts and mailbox posts or give it a trellis with allamandas and pink trumpet vines. Make it a centerpiece with bat face cupheas, coleus, taro, and yellow bells nearby.

Try These
M. sanderi, Brazilian jasmine, has cultivars with white flowers and yellow throats.

Native to Central and South America, the rosy pink trumpets of mandevilla vines call out to passersby to put on the brakes and simply enjoy their beauty. Much of the appeal of tropical plants outside of frost-free areas is embodied by their ability to get people to slow down the pace of life and drift away, if only for a moment. Spend that moment watching a bumblebee climbing into its flower tube. Be amazed at the hot pink blossoms popping open along shiny, dark green leaves that charge up vigorously and grab their supports. Grow any mandevilla available; they are all kin and the result of both natural and induced hybridization. For example, Mandevilla amoena 'Alice du Pont' is a larger vine with smaller leaves and flowers in clusters. The plants are well behaved and equally at home flanking a swimming pool or spilling from a basket indoors and out.

Plant Family
Apocynaceae

Other Common Name
Pink allamanda

Bloom Period and Seasonal Color
Pink, red, or white flowers spring through the fall

Mature Height × Spread
Up to 10 ft. vine

Mexican Flame Vine

Pseudogynoxys chenopodioides

Native to Mexico and as colorful as a sunset sky, Mexican flame vine is one hot plant! The vines make dense shade, covered in dark green leaves that are both succulent and toothy. Fast-growing flame vine easily covers any structure available and will bloom equally well growing up a trellis, down a wall, or across a pergola. The flowers are screaming orange lollipops borne in clusters that last for weeks and eventually turn bright red. Their shape falls somewhere between Mexican zinnia and a big coneflower with a clown nose. They are like a Dr. Seuss version of a classic daisy, long, radiant petals around a center that is a fancy frilled knob. Mexican flame vines, with a blanket of thick, lush leaves decorated with blinding orange flowers, bloom year-round in the tropics and often become perennial in the warm subtropics.

Plant Family
Asteraceae/Compositae

Other Common Name
None

Bloom Period and Seasonal Color
Orange and orange-red flowers with yellow centers in the summer and fall

Mature Height × Spread
Up to 30 ft. vine; averages 12 ft.

When, Where, and How to Plant

Select a sunny space with a reliable water source nearby to meet flame vine's great needs for water. These plants can grow well outdoors in the tropics and warm subtropics as perennials and are easily grown as annual vines elsewhere. Mexican flame plants are not drought tolerant but are otherwise easy to grow. They thrive in a soil that can be watered regularly without becoming saturated. Amend existing garden soils and potting mixes with organic matters to create a soil that is richly organic, fertile, and well drained. Space plants 1 foot apart along a sturdy support in beds, or provide a pot 12 inches deep and wide for one plant with a 6- to 8-foot-tall trellis. Mulch beds and pots 2 inches deep with organic material such as shredded or ground bark.

Growing Tips

Water Mexican flame vines very regularly. Let the soil dry out only a little between irrigations and be sure to water before fertilizing. Use a general-purpose garden fertilizer regularly, either a granular or soluble formula. Train young vines in a vertical pattern while they are supple. Work mulch into the soil as it rots and replenish it promptly. Let flowers dry on the vine in late summer in order for seedpods to develop.

Care and Propagation

Few pests "bug" Mexican flame vines. Propagate them by cuttings taken at midsummer and from seeds you saved.

Companion Planting and Design

Such bright orange color needs bright and dark bedmates such as copperleafs, taro, coleus, and Chinese hat plants. Grow a row of tropical vines with hyacinth beans, mandevillas, and Chinese violets for striking, fast shade.

Try These

For similar bright orange in bedding plants, grow *Tithonia*, Mexican sunflower.

Moonflower
Ipomoea alba

When, Where, and How to Plant

Moonflower vines thrive in garden soil in warm weather everywhere, produce viable seed, and often reseed in the subtropics. Give the vines space to grow by providing a support at least 4 feet wide and 8 feet tall. Choose a sunny site with a reliable water source close at hand. Prepare a soil that is organic, fertile, and well drained by amending existing garden soil with organic matters. Plant seed or space plants 6 inches apart in a row below the trellis and if needed, tie strings to the trellis that reach to the vines for a fast grab. Mulch beds of moonflower with 2 inches of ground or shredded bark. Keep new plantings very well watered until they are up and growing.

Growing Tips

Water moonflower plants deeply and regularly and let them dry out only slightly between irrigations. Use a general-purpose garden fertilizer very regularly, as often as directed from the spring through fall, to ensure vigorous vines. Train young vines to the support you give them at the first opportunity, while they are pliable. Maintain organic mulch around the plants and work it into the soil as it decomposes. If you let some flowers mature, they will form seedpods. Dry them on the vine, if possible, and store them dry (not frozen).

Care and Propagation

Watch for chewing caterpillars on young plants. Propagate moonflowers by seeds if you want to expand your collection.

Companion Planting and Design

It is hard to beat moonflowers for fast shade and night garden excitement. Grow some with night-blooming jasmines and night-blooming cereus nearby or with morning glories for flowers day and night.

Try These

Ipomoea lobata is called Spanish flag vine; it has tight chains of vivid yellow and red flowers.

Native to the American tropics, moonflower vines deliver massive leaves both in size and numbers grand enough to be a tropical standout. Heart-shaped and abundant, they are thick, slightly waffled, and densely packed on their vines to create cooling shade. The thick, fast-growing vines soon toss out flower tubes everywhere, 6 inches long and whiter than white. They look like candles awaiting the flame. The blooms open to great fanfare when night falls and quickly fill the air with their sultry perfume. These ladies of the night are boisterous and bawdy when seen as a group, yet a single flower could be a centerpiece. As big as saucers and deeply veined in cream, each shines like sateen and has a coquettish ruffle at the edges. To view them after dark aflutter with visiting moths is one of the truest pleasures of being a gardener.

Plant Family
Convolvulaceae

Other Common Name
Moon vine

Bloom Period and Seasonal Color
White flowers in the summer and fall

Mature Height × Spread
Up to 20 ft. vine; averages 12 ft.

Passion Flower
Passiflora alata

Passion flower evokes different images to different people. Some see the religious imagery in the arrangements of the stamens, while others say the flowers are passionate shades of purple and red. Either way, this is a captivating vine, scrambling fast up and over any handy support. Each leaf is a wavy, almost dainty-looking teardrop shape, with strongly marked veins and slightly scalloped edge. The flowers of this species are stacked like a fancy dessert torte. At the base are long red petals arranged like a plate to hold the next level, a sunburst of long thin purple fringe. From its center, the waxy yellow and white styles and stigmas pop up like tricorn hats. Their fragrance is darkly sweet and musky to fulfill its passionate reputation. Passion flower draws its other common name from yellow-orange egg-shaped fruit called a granadilla in its native Brazil.

Plant Family
Passifloraceae

Other Common Name
Fragrant granadilla

Bloom Period and Seasonal Color
Red-white flowers with red inner summer orange fruits

Mature Height × Spread
Up to 20 ft. vine; averages 8 ft.

When, Where, and How to Plant
Choose a sunny site with a reliable water source nearby. Passion flower vines thrive in fertile, organic, well-drained soil. Amend existing garden soils and potting mixes with compost, ground barks, and other aged products such as manure if needed to achieve these conditions. Start seeds indoors in the winter to transplant into the garden in warm weather. Provide a support pole or trellis at least 8 feet tall and be prepared to guide the vine as it grows. Tender young sprouts are easily interwoven and will bloom most abundantly when they are growing horizontally or hanging down. Give one plant a pot 10 inches wide and as deep, or space vines 1 foot apart along a pergola or fence. Mulch with 2 inches of organic material such as shredded bark.

Growing Tips
Regular watering is essential to healthy vines and blooms, and even after flowering if fruit is desired. Water the plants deeply and allow the soil to dry out only slightly before watering again. Fertilize regularly with a complete formula product if growing passion flower as an annual. Fertilize established vines three or four times annually. As the mulch rots, work it into the soil and replace it promptly. Take cuttings in late summer for young plants to carry over the winter, or harvest seedpods.

Care and Propagation
Passion flowers have few pests, but the gulf fritillary and many other desirable butterfly caterpillars may defoliate the plant. Propagate by cuttings and seed.

Companion Planting and Design
Let passion flowers climb high with a group below such as red gingers, butterfly flowers, and fragrant heliotrope. Grow a group of tropical vines with star jasmines, Philippine orchids, and flaming glorybowers.

Try These
'Ruby Glow' looks like its name, with flowers that are stunning shades of red and purple.

Pink Trumpet Vine

Podranea ricasoliana

When, Where, and How to Plant

Choose a site in sun or mostly sun with a reliable water source nearby. Provide a soil to grow pink trumpet vine that is rich, organic, and fertile with very good drainage. Amend existing garden soils and potting mixes with compost and other organic matters to create these conditions where they do not exist. Set up a sturdy trellis and space plants 10 to 12 inches apart in beds and planters. Lay strings from the base of the trellis to the vines to guide the vigorous growth onto their supports. The large vines and their rooted cuttings can be maintained over the winter as container plants in the temperate zone. Mulch both beds and pots with 2 inches of organic material such as ground bark or pine straw.

Growing Tips

Water pink trumpet vine plants regularly and let the soil dry out only slightly between irrigations. Use a general-purpose garden fertilizer as often as directed from the spring through fall, but less often in the winter. Work rotted mulch into the soil and replace it promptly with fresh organic material. Prune vines back by one-third in late fall outdoors and in late winter for potted plants to control height and stimulate new vines.

Care and Propagation

Few pests bother pink trumpet vines, but susceptibility to nematodes in Florida necessitates planting in sterile soil. Propagate by tip cuttings and seed.

Companion Planting and Design

Plant a collection of sun-loving vines including pink trumpets, golden trumpets, Mexican flame vines, and Chinese violets. Grow pink trumpet vines with Persian shield, cat's whiskers, and night-blooming jasmines.

Try These

Queen of Sheba vine (*P. brycei*) is a smaller vine with yellow-throated flowers.

Pink trumpet vine could easily be loved for its masses of narrow leaves with nifty zigzag edges that cover an arbor in no time. The vines grow into a dense blanket of foliage, the perfect backdrop for their papery flowers, rather like how a matte green border sets off the oil painting it surrounds. The twining vines are fast growing and easily trained while their new growth is tender. Clusters of white-pink buds crowd the stem tips and blast open into wide, flat-faced pink trumpets that look more like French horns than cornets in the way their bells flare out. Snappy pink in color with a deceptively ethereal appearance, the flowers are distinguished by thin, dark red veins that paint stripes near the centers, reminiscent of some azaleas. Native to South Africa, pink trumpet vine can be made to feel at home anywhere warm temperatures, good air circulation, and bright sunlight prevail.

Plant Family
Bignoniaceae

Other Common Name
Port St. John's creeper

Bloom Period and Seasonal Color
Pink flowers with red veins and rose throats in the summer

Mature Height × Spread
Up to 20 ft. vine; usually maintained at 10 ft.

Scrambling Clock Vine
Thunbergia battiscombei

Scrambling clock vine is native to tropical Africa and a seasoned traveler with no reservations, ready to encamp anywhere. Without a trellis or plant to support it, the vines collapse neatly into a green mound of leaves and cornflower blue flowers. It is not aggressive and grows well in containers, but it is opportunistic and will climb over whatever is handy in search of sunlight needed to produce the rich flowers. The leaves are shiny, light green, and heart-shaped, as close to dainty as a tropical leaf can be. The flower clusters, called racemes, appear in the leaf axils, the place where the stem and leaf meet. Their floriferous habit means the plants are covered in blossoms for months even in the temperate zone. Each flower begins as a fuzzy white bud that elongates into a tube and finally flares open with bright bluish purple trumpets.

Plant Family
Acanthaceae

Other Common Name
Scrambling sky vine

Bloom Period and Seasonal Color
Deep blue flowers with yellow throats appear from the spring through fall

Mature Height × Spread
Up to 6 ft. × 2 ft.

When, Where, and How to Plant
Choose a sunny site or one that is mostly sunny for this outstanding small vine. It can grow outdoors all year in the subtropics and tropics, and young plants do well in warm, bright rooms indoors. Scrambling clock vine is not drought tolerant but it cannot stand a saturated root zone either. It requires a well-drained, organic, fertile soil in beds or pots. Prepare that kind of growing medium by mixing garden soil or potting mix with finely ground bark or sand, compost, and other organic matters. Plant young vines 12 inches apart along a 6-foot-tall trellis or fence, or provide a container 8 to 10 inches wide and deep with a teepee or column to support the plants. Make sure new plantings do not dry out before they become established.

Growing Tips
Keep this plant on a regular schedule that allows the plants to dry out between waterings, but not excessively so. Begin fertilizing when new growth emerges. Apply a complete formula, all-purpose fertilizer regularly during the growing season. Use a granular garden product or its soluble equivalent. If plant growth slows or it does not flower, water more often. Limit mulch to 1 inch of organic matter and work it into the bed as it decomposes.

Care and Propagation
There are very few pests that will bug scrambling clock vines. Propagate more plants by cuttings taken in summer or grow from seed in late winter.

Companion Planting and Design
Grow scrambling clock vines with foxtail agaves or spineless yuccas for contrast. Combine them in pots with kanga paw and ti plants for color and texture. Train one up or let it sprawl on a fence with corkscrew vine.

Try These
T. grandiflora has lavender blue hues, while *T. mysorensis* is red and yellow.

Windowleaf

Monstera deliciosa

When, Where, and How to Plant

Prepare a garden soil or container growing mix that is richly organic yet well drained. For windowleaf plants to thrive, the soil must be consistently damp. Amendments to available soils to achieve these conditions might include ground bark, compost, and other organic matters. Select a site outdoors in shade or bright light without direct sun for garden plantings and containers. Indoors, choose a spot in a bright room away from sunny windows. Space young plants 2 feet apart in beds or provide a container 12 inches deep and wide for one windowleaf. Water new plantings well and provide a column or tree to support the plants. Mulch beds and pots with 2 inches of an organic material such as ground or shredded bark.

Growing Tips

Water very regularly supplied is necessary to grow and maintain windowleaf, especially in its first years. The plants are never drought tolerant but do not survive wet feet either. Add a soluble, general-purpose fertilizer to the water monthly for the first year, then every other month. Work mulch into the soil when it rots and replace it promptly. Repot windowleafs annually or when you spot roots growing through the drain hole.

Care and Propagation

Watch for mealybugs, but that's about it for pests. Propagate by cuttings to expand your collection. See "Propagation" (page 36) for details.

Companion Planting and Design

Grow a green screen with windowleafs, ruffle palms, elephant ears, and African masks. Accent windowleaf plants with colorful golden brush gingers, nerve plants, blushing bromeliads, and white bat flowers.

Try These

Cultivars with wildly patterned leaves include 'Variegata' and 'Albovariegata', which has white markings.

Windowleaf is a huge comfort plant— the macaroni and cheese of the tropical garden—warm, substantial, and easy to love. Bawdy and bold with leaves as big as chair cushions, it is a creeping, viney thing tough enough to survive low light and humidity. Conveniently, windowleaf will grow a top in proportion to its available root space and thus adapts readily to long life in large containers. Each leaf is deeply cut to the midrib, creating fat ribbons of glossy green. With age, holes develop within the ribbons and give windowleaf another of its common names, Swiss cheese plant. The stout vines grab tree trunks in their native rainforests, throwing out aerial roots like so many lifelines to the floor below as they climb. Classic arum family flowers arise on fleshy stems, one big-cupped bract lovingly wrapped around a long tube of blossom. Some forms produce an edible fruit that is popular in the tropics.

Plant Family
Araceae

Other Common Name
Swiss cheese plant

Bloom Period and Seasonal Color
Evergreen leaves

Mature Height × Spread
Up to 20 ft. × 10 ft.; averages 6 ft. × 3 ft.

Plants for Special Sites

Shady dry: Let the soil dry out between waterings. Wait to water again until the soil feels dry up to the first knuckle of your index finger. Many of these can also grow well in part sun.

Chamaedorea seifrizii	bamboo palm
Clivia miniata	bush lily
Codiaeum variegatum pictum	croton
Crassula ovata	jade plant
Cryptanthus hybrid	earth star
Dizygotheca elegantissima	false aralia
Epiphyllum oxypetalum	queen of the night
Ficus benjamina	weeping fig
Kalanchoe fedtschenkoi	South American air plant
Peperomia obtusifolia	blunt-leaf peperomia
Sansevieria trifasciata	snake plant
Schefflera arboricola	dwarf schefflera

Shady wet: Let the soil dry out slightly between waterings.

Anthurium spp.	flamingo flower
Asplenium nidus	birds nest fern
Begonia × *semperflorens-cultorum*	wax begonia
Begonia hybrids	angel wing begonia
Begonia rex	Rex begonia
Begonia spp.	rhizomatous begonia
Brunfelsia pauciflora	yesterday, today, and tomorrow
Caladium bicolor	caladium
Calathea lancifolia	rattlesnake plant
Calathea makoyana	peacock plant
Dieffenbachia hybrids	dieffenbachia
Impatiens walleriana	impatiens
Justicia spp.	jacobinia
Medinilla myriantha	Malaysian orchid
Neoregelia carolinae	striped blushing bromeliad
Phaius tankervilleae	nun's orchid
Philodendron selloum	cut-leaf philodendron
Pseuderanthemum alatum	chocolate plant

Rhapis excelsa	lady palm
Strobilanthes dyeranus	Persian shield

Shady wet consistently: Do not let soil dry out, maintain damp soil.

Achimenes spp.	orchid pansy
Aiphanes horrida	ruffle palm
Alocasia × amazonica	African mask
Alocasia odora	upright elephant ear
Burbidgea schizocheila	golden brush ginger
Curcuma spp.	hidden ginger
Fittonia albivinis	nerve plant
Maranta spp.	prayer plant
Monstera deliciosa	windowleaf
Nepenthes alata	monkey cup
Nephrolepis spp.	fern
Platycerium spp.	staghorn fern
Pogostemon cablin	patchouli
Tacca integrifolia	white bat flower
Xanthosoma sagittifolium	elephant ear

Sunny dry: Let the soil dry out between waterings. Wait to water again until the soil feels dry up to the first knuckle of your index finger.

Agave attenuata	foxtail agave
Ananas comosus	edible pineapple
Calliandra haematocephala 'Nana'	dragon powder puff
Callistemon viminalis	bottlebrush
Callistephus chinensis	China aster
Clerodendrum incisum	musical note plant
Congea tomentosum	shower orchid
Cordyline fruticosa (terminalis)	ti plant
Costus woodsonii	dwarf cone ginger
Crossandra infundibuliformis	firecracker flower
Draceana draco	dragon tree
Elaeocarpus grandiflorus	fairy petticoat
Euphorbia tithymaloides	devil's backbone

Plants for Special Sites *(continued)*

Euryops chrysanthemoides	Paris daisy
Gardenia carinata	yellow gardenia
Gazania rigens	gazania
Gerbera spp.	gerber daisy
Hibiscus rosa-sinensis	hibiscus
Hoya carnosa	wax plant
Ixora coccinea	flame of the woods
Leonotis leorunus	lion's ear
Malope trifida	mallow
Malpighia glabra	Barbados cherry
Osteospermum ecklonis	Cape daisy
Phoenix roebelenii	miniature date palm
Pimenta dioica	allspice
Russelia equisetiformis	firecracker plant
Thunbergia battiscombei	scrambling clock flower
Vigna caracalla	corkscrew flower
Yucca elephantipes	spineless yucca

Sunny dry consistently: Mature plants will be drought tolerant.

Adenium obesum	desert rose
Aloe ferox	Cape aloe
Bougainvillea spp.	bougainvillea
Euphorbia milii	crown of thorns
Euphorbia umbellate	African milk bush

Sunny wet: Let soil dry out slightly between waterings.

Acalypha hispida	chenille plant
Acalypha wilkesiana	copperleaf
Allamanda cathartica	golden trumpet
Angelonia angustifolia	summer snapdragon
Anigozanthos flavidus	kangaroo paw
Arundina graminifolia	bamboo orchid
Brugmansia suaveolens	angel trumpet
Bunchosia argentea	peanut butter fruit
Cananga odorata	dwarf ylang ylang

Plants for Special Sites *(continued)*

Cestrum aurantiacum	yellow cestrum
Cestrum nocturnum	night-blooming jasmine
Cinnamomum verum	cinnamon tree
Citrus aurantifolia	key lime
Citrus sinensis	navel orange
Coleus × hybridus	coleus
Colocasia esculenta	taro
Costus spicatus	dwarf cone ginger
Cuphea llavea	bat face cuphea
Cycas revoluta	sago palm
Datura metel	horn of plenty
Dolichos purpureus	hyacinth bean vine
Duranta erecta	golden dewdrop
Ensete ventricosum	red Abyssinian banana
Epidendrum spp.	reed stem orchid
Gloriosa superba	glory lily
Graptophyllum pictum	caricature plant
Helichrysum petiolare	licorice plant
Holmskioldia sanguinea	Chinese hat plant
Hypoestes phyllostachya	polka dot plant
Ipomoea alba	moonflower
Jatropha multifida	coral plant
Magnolia champaca	joy perfume tree
Mandevilla spp.	mandevilla vine
*Megaskepasma erythrochlamys*red	justicia
Mitriostigma axillare	African gardenia
Murraya koenigii	curry leaf
Orthosiphon aristatus	cat's whiskers
Pachystachys lutea	lollipop flower
Pereskia grandifolia	rose cactus
Plumbago auriculata	skyflower
Plumeria hybrids	frangipani
Podranea ricasoliana	pink trumpet vine
Salvia coccinea	scarlet sage

Plants for Special Sites *(continued)*

Scaevola aemula	fan flower
Senecio confusus	Mexican flame vine
Senna alata	candle bush
Stephanotis floribunda	Madagascar jasmine
Strelitzia reginae	bird of paradise
Tecoma stans	yellow bells
Telosma cordata	Chinese violet

Sunny wet consistently: Do not let soil dry out, maintain damp soil.

Alpinia purpurata	red ginger
Alternanthera brasiliana	calico plant
Carica papaya	papaya
Clerodendrum splendens	flaming glorybower
Cyperus alternifolius	umbrella plant
Freycinetia cumingiana	flowering pandanus
Heliconia stricta	lobster claw
Heliotropium arborescens	fragrant heliotrope
Jasminum laurifolium var. *laurifolium*	angelwing jasmine
Jasminum sambac	Arabian jasmine
Pachira aquatica	money tree
Passiflora alata	passion flower
Sanchezia speciosa	sanchezia
Spathoglottis plicata	Philippine orchid
Strongylodon macrobotrys	emerald creeper
Tabernaemontana divaricata	carnation of India

Plants for Special Features

Evergreen Leaves

Agave attenuata	foxtail agave
Aiphanes horrida	ruffle palm
Asplenium nidus	birds nest fern
Carica papaya	papaya
Chamaedorea seifrizii	bamboo palm
Colocasia esculenta	taro
Crassula argentea	jade plant
Cycas revoluta	sago palm
Dieffenbachia hybrids	dumb cane
Dizygotheca elegantissima	false aralia
Draceana draco	dragon tree
Ficus benjamina	weeping fig
Monstera deliciosa	windowleaf
Nephrolepis spp.	fern
Peperomia obtusifolia	blunt-leaf peperomia
Philodendron selloum	cut-leaf philodendron
Phoenix roebelenii	miniature date palm
Platycerium spp.	staghorn fern
Rhapis excelsa	lady palm
Sansevieria trifasciata	snake plant
Schefflera arboricola	dwarf schefflera
Strongylodon macrobotrys	emerald creeper
Xanthosoma sagittifolium	elephant ear
Yucca elephantipes	spineless yucca

Spectacular Flowers

Acalypha hispida	chenille plant
Achimenes spp.	orchid pansy
Aechmea fasciata	silver vase bromeliad
Allamanda cathartica	golden trumpet
Angelonia angustifolia	summer snapdragon
Anigozanthos flavidus	kangaroo paw
Anthurium spp.	flamingo flower

Plants for Special Features *(continued)*

Arundina graminifolia	bamboo orchid
Begonia × semperflorens-cultorum	wax begonia
Brugmansia suaveolens	angel trumpet
Brunfelsia pauciflora	yesterday, today, and tomorrow
Calliandra haematocephala 'Nana'	dragon powder puff
Callistemon viminalis	bottlebrush
Callistephus chinensis	China aster
Cestrum aurantiacum	yellow cestrum
Clivia miniata	bush lily
Crossandra infundibuliformis	firecracker flower
Cryptanthus hybrid	earth star
Cuphea llavea	bat face cuphea
Datura metel	horn of plenty
Duranta erecta	golden dewdrop
Epidendrum spp.	reed stem orchid
Epiphyllum oxypetalum	queen of the night
Euphorbia umbellate	African milk bush
Euryops chrysanthemoides	Paris daisy
Gazania rigens	gazania
Gerbera spp.	gerber daisy
Heliconia stricta	lobster claw
Hibiscus rosa-sinensis	hibiscus
Holmskioldia sanguinea	Chinese hat plant
Hypoestes phyllostachya	polka dot plant
Impatiens walleriana	impatiens
Ixora coccinea	flame of the woods
Jatropha multifida	coral plant
Justicia spp.	jacobinia
Kalanchoe fedtschenkoi	South American air plant
Leonotis leorunus	lion's ear
Malope trifida	mallow
Malpighia glabra	Barbados cherry
Mandevilla spp.	mandevilla vine
Medinilla myriantha	Malaysian orchid

Plants for Special Features *(continued)*

Megaskepasma erythrochlamys	Brazilian red orchid
Mitriostigma axillare	African gardenia
Nepenthes alata	monkey cup
Osteospermum ecklonis	Cape daisy
Pachystachys lutea	lollipop flower
Pereskia grandifolia	rose cactus
Phaius tankervilleae	nun's orchid
Plumbago auriculata	skyflower
Podranea ricasoliana	pink trumpet vine
Rotheca myriocoides	blue butterfly flower
Salvia coccinea	scarlet sage
Scaevola aemula	fan flower
Senecio confusus	Mexican flame vine
Spathoglottis plicata	Philippine orchid
Strelitzia reginae	bird of paradise
Tecoma stans	yellow bells
Thunbergia battiscombei	scrambling clock vine
Vigna caracalla	corkscrew flower

Amazing Painted Leaves

Acalypha wilkesiana	copperleaf
Alocasia × amazonica	African mask
Begonia rex	Rex begonia
Caladium bicolor	caladium
Calathea lancifolia	rattlesnake plant
Calathea makoyana	peacock plant
Codiaeum variegatum pictum	croton
Coleus × hybridus	coleus
Cordyline fruticosa (terminalis)	ti plant
Ensete ventricosum	red Abyssinian banana
Euphorbia tithymaloides	devil's backbone
Fittonia albivinis	nerve plant
Graptophyllum pictum	caricature plant
Helichrysum petiolare	licorice plant

Plants for Special Features *(continued)*

Maranta spp.	prayer plant
Neoregelia carolinae	striped blushing bromeliad
Sanchezia speciosa	sanchezia
Strobilanthes dyeranus	Persian shield

Extraordinary Fragrance in Leaves

Ananas comusus	edible pineapple
Bunchosia argentea	peanut butter fruit
Cinnamomum verum	cinnamon tree
Citrus aurantifolia	key lime
Citrus sinensis	navel orange
Murraya koenigii	curry leaf

Extraordinary Fragrance in Flowers

Cestrum nocturnum	night-blooming jasmine
Clerodendrum splendens	flaming glorybower
Congea tomentosum	shower orchid
Elaeocarpus grandiflorus	fairy petticoat
Freycinetia cumingiana	flowering pandanus
Gardenia carinata	yellow gardenia
Gloriosa superba	glory lily
Heliotropium arborescens	fragrant heliotrope
Hoya carnosa	wax plant
Ipomoea alba	moonflower
Jasminum laurifolium	angel wing jasmine
Jasminum sambac	Arabian jasmine
Michelia champaca	joy perfume tree
Pachira aquatica	money tree
Passiflora alata	passion flower
Plumeria hybrids	frangipani
Pogostemon cablin	patchouli
Stephanotis floribunda	Madagascar jasmine
Tabernaemontana divaricata	carnation of India
Tacca integrifolia	white bat flower
Telosma cordata	Chinese violet

Plants for Special Features *(continued)*

Drought Tolerance

Adenium obesum	desert rose
Aloe ferox	Cape aloe
Bougainvillea spp.	bougainvillea
Euphorbia milii	crown of thorns
Euphorbia umbellate	African milk bush

Water Garden Plants

Alocasia odora	upright elephant ear
Alpinia purpurata	red ginger
Alternanthera brasiliana	calico plant
Burbidgea schizocheila	golden brush ginger
Curcuma spp.	hidden ginger
Cyperus alternifolius	umbrella plant

Resources

Resources and Helpful Information

Books

Bawden-Davis, Julie. *Indoor Gardening the Organic Way.* Lanham, MD: Taylor Trade Publishing, 2007.

Coombes, Allen, ed. *The Timber Press Dictionary of Plant Names.* Portland, OR: Timber Press, 2009.

Foster, Lee. *Gardening Techniques.* San Francisco: Ortho Books, 1985.

Lawton, Barbara Perry. *Hibiscus: Hardy and Tropical Plants for the Garden.* Portland, OR: Timber Press, 2004.

Pleasant, Barbara. *Warm Climate Gardening.* Pownal, VT: Storey Communications, 1993.

Roth, Sally. *Attracting Butterflies and Hummingbirds to Your Backyard.* Emmaus, PA: Rodale Press, 2001.

Riffle, Robert Lee. *The Tropical Look: An Encyclopedia of Dramatic Landscape Plants.* Portland, OR: Timber Press, 1998.

Thompson, Peter. *Creative Propagation: A Grower's Guide.* Portland, OR: Timber Press , 1992.

———. *The Looking-Glass Garden: Plants and Gardens of the Southern Hemisphere.* Portland, OR: Timber Press, 2001.

Van Aken, Norman. *The Great Exotic Fruit Book.* Berkeley, CA: Ten Speed Press, 1995.

White, Judy. *Taylor's Guide to Orchids.* Boston: Houghton Mifflin, 1996.

Winters, Norman. *Paradise Found: Growing Tropicals in Your Own Backyard.* Dallas, TX: Taylor Trade Publishing, 2001.

Websites

www.stokestropicals.com

www.logees.com

www.arborday.org

www.soilfoodweb.com.

www.floridata.com

www.therainforestgarden.com

Photo Credits

Cool Springs Press would like to thank the following photographers and illustrators for their contributions to *Gardener's Guide to Tropical Plants*.

Photo by P. Acevado courtesy of Smithsonian Institution: Page 158

Bill Adams: Page 120

James Anglin: Pages 7, 9, 43, 47

Steve Asbell: Pages 6, 8, 74, 80, 83, 84, 91, 93, 100, 103, 104, 108, 111, 128, 129, 136, 143, 145, 150, 151, 154, 155, 156, 159, 164, 166, 168, 187, 203, 207

Liz Ball: Page 162

Cathy Barash: Page 16

Bidgeee/Commons.wikimedia.org: Page 28

Botanic Images Inc./Garden World Images: Page 140

Cpoddwrites/Commons.wikimedia.org: Page 29

Gilles Delacroix/Garden World Images: Pages 125, 138

Tom Eltzroth: Pages 13, 15, 18, 40, 45, 48, 49, 60, 67, 72, 86, 88, 96, 99, 112, 115, 116, 118, 119, 122, 124, 126, 147, 152, 157, 165, 172, 173, 174, 177, 178, 179, 181, 184, 189, 192, 198, 200

Katie Elzer-Peters: Pages 22, 37, 38, 63, 82, 135

Joyce Grigonis: Page 134

Lorenzo Gunn: Page 186

Stephen Hauser/Garden World Images: Pages 109, 163

Richard A. Howard Image Collection, courtesy of Smithsonian Institution: Pages 117, 148, 170, 193, 202

iStockphoto: Pages 65, 127, 204

Jenny Lilly/Garden World Images: Page 139

MAP/Nicole et Patrick Mioulane/Garden World Images: Page 196

Jerry Pavia: Pages 10, 41, 44, 61, 89, 98, 102, 137, 146, 153, 175, 176, 182, 199

Shutterstock: Pages 11, 12, 14, 17, 21, 27, 31, 32, 33, 46, 51, 56, 57, 58, 59, 62, 66, 69, 70, 73, 76-77, 101, 110, 113, 131, 149, 160-161, 169, 183, 190-191

Neil Soderstrom: Page 36

Brad Springer: Page 55

Forest and Kim Starr: Page 185

Lynn Steiner: Page 30

Glenn Stokes (www.stokestropicals.plants.com): Pages 26, 42, 50, 78, 79, 81, 85, 87, 90, 92, 94, 95, 97, 105, 106, 107, 114, 121, 123, 130, 132, 133, 141, 142, 144, 167, 171, 180, 188, 195, 196, 197, 205, 206

Andre Viette: Page 201

Barbara Wise: Page 53

USDA Hardiness Zone Map

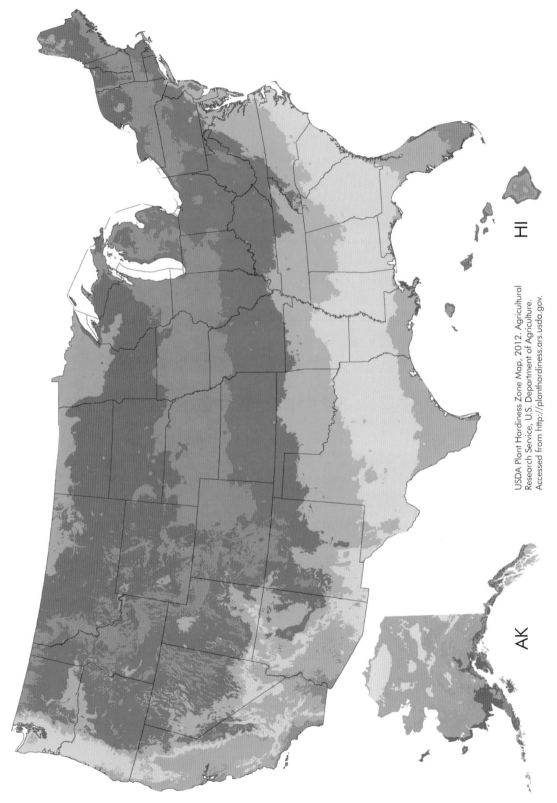

HI

AK

USDA Plant Hardiness Zone Map, 2012. Agricultural
Research Service, U.S. Department of Agriculture.
Accessed from http://planthardiness.ars.usda.gov.

Average Annual Extreme Miniature Temperature 1976–2005

Temp (F)	Zone	Temp (C)	Temp (F)	Zone	Temp (C)
-60 to -55	1a	-51.1 to -48.3	5 to 10	7b	-15 to -12.2
-55 to -50	1b	-48.3 to -45.6	10 to 15	8a	-12.2 to -9.4
-50 to -45	2a	-45.6 to -42.8	15 to 20	8b	-9.4 to -6.7
-45 to -40	2b	-42.8 to -40	20 to 25	9a	-6.7 to -3.9
-40 to -35	3a	-40 to -37.2	25 to 30	9b	-3.9 to -1.1
-35 to -30	3b	-37.2 to -34.4	30 to 35	10a	-1.1 to 1.7
-30 to -25	4a	-34.4 to -31.7	35 to 40	10b	1.7 to 4.4
-25 to -20	4b	-31.7 to -28.9	40 to 45	11a	4.4 to 7.2
-20 to -15	5a	-28.9 to -26.1	45 to 50	11b	7.2 to 10
-15 to -10	5b	-26.1 to -23.3	50 to 55	12a	10 to 12.8
-10 to -5	6a	-23.3 to -20.6	55 to 60	12b	12.8 to 15.6
-5 to 0	6b	-20.6 to -17.8	60 to 65	13a	15.6 to 18.3
0 to 5	7a	-17.8 to -15	65 to 70	13b	18.3 to 21.1

Arbor Day Map

Differences between 1990 USDA hardiness zones and 2006 arborday.org hardiness zones reflect warmer climate

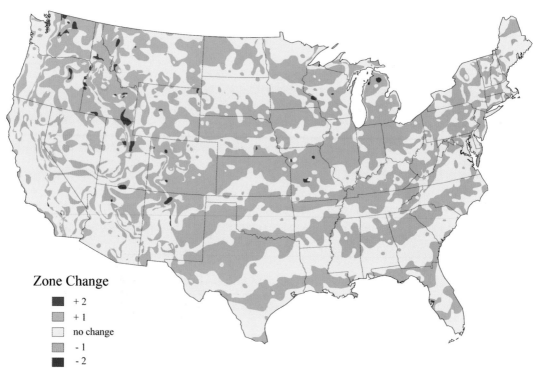

Zone Change

- ■ + 2
- ▨ + 1
- ☐ no change
- ▨ - 1
- ■ - 2

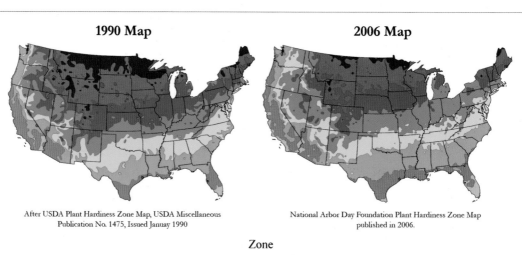

1990 Map

After USDA Plant Hardiness Zone Map, USDA Miscellaneous
Publication No. 1475, Issued Januay 1990

2006 Map

National Arbor Day Foundation Plant Hardiness Zone Map
published in 2006.

Zone

2 3 4 5 6 7 8 9 10

© 2006 by The National Arbor Day Foundation®

Plant Index

Page numbers in italics indicate photographs.

Meet Nellie Neal

Known as "The Garden Mama" to her radio audiences, Nellie Neal writes and speaks from a solid educational background coupled with a lifetime of gardening successes and failures. She learned to garden from her grandfather in Monroe, Louisiana, and majored in English and Horticulture at LSU in Baton Rouge. In her long career, Neal has grown plants, bought and sold plants, tested new varieties of plants, and helped sustain old ones. She has planted and maintained landscapes and movie sets, mowed lawns, watered greenhouses, waited on customers in garden retail, and taught gardening to students of every age.

Nellie began writing and speaking about gardening from her home office in 1990 and began her radio programs in 1994. She is a garden advocate who presents formal lectures, demonstrations, and workshops to twenty groups annually. Neal posts regional reports online at www.nationalgardening.com, where she writes twice monthly about the southern coastal and tropics regions. She blogs weekly at her website, **gardenmama.com**, about her garden and her propagation passion. She tweets organic gardening tips and writes question-and-answer columns for the *Clarion Ledger* newspaper and *Mississippi Gardener* magazine. The owner of GardenMama, Inc., Neal is a member of the Garden Writers Association, a founding member of **greatgardenspeakers.com**, and a member of IATSE 478. She is the author of several books including *Questions and Answers for Deep South Gardeners*, 1st and 2nd Editions; *Ortho's All About Greenhouses*; *Ortho's All About Houseplants*; *Organic Gardening Down South*; and *How to Get Started in Southern Gardening* for Cool Springs Press.

Nellie and her husband, Dave Ingram, have four adult children, three cats, and one beautiful acre in the historic Fondren neighborhood of Jackson, Mississippi, where they grow plants and vegetables and they devote time to two city gardens in New Orleans. Nellie Neal has never met a plant she didn't want to propagate.

Gardening Notes

Gardening Notes

Gardening Notes

Gardening Notes

Gardening Notes

Gardening Notes